ENHANCING THE EFFECTIVENESS OF
TEAM SCIENCE

D1457995

Nancy J. Cooke and Margaret L. Hilton, *Editors*

Committee on the Science of Team Science

Board on Behavioral, Cognitive, and Sensory Sciences

Division of Behavioral and Social Sciences and Education

NATIONAL RESEARCH COUNCIL
OF THE NATIONAL ACADEMIES

THE NATIONAL ACADEMIES PRESS
Washington, D.C.
www.nap.edu

THE NATIONAL ACADEMIES PRESS 500 Fifth Street, NW Washington, DC 20001

This study was supported by the National Science Foundation under Contract No. OCI-1248170 and by Elsevier. Any opinions, findings, conclusions, or recommendations expressed in this publication are those of the authors and do not necessarily reflect the views of the organizations or agencies that provided support for the project.

International Standard Book Number-13: 978-0-309-31682-8
International Standard Book Number-10: 0-309-31682-0
Library of Congress Control Number: 2015940916

Additional copies of this report are available from the National Academies Press, 500 Fifth Street, NW, Keck 3607, Washington, DC 20001; (800) 624-6242 or (202) 334-3313; http://www.nap.edu.

Suggested citation: National Research Council. (2015). *Enhancing the Effectiveness of Team Science*. Committee on the Science of Team Science, N.J. Cooke and M.L. Hilton, Editors. Board on Behavioral, Cognitive, and Sensory Sciences, Division of Behavioral and Social Sciences and Education. Washington, DC: The National Academies Press.

THE NATIONAL ACADEMIES
Advisers to the Nation on Science, Engineering, and Medicine

The **National Academy of Sciences** is a private, nonprofit, self-perpetuating society of distinguished scholars engaged in scientific and engineering research, dedicated to the furtherance of science and technology and to their use for the general welfare. Upon the authority of the charter granted to it by the Congress in 1863, the Academy has a mandate that requires it to advise the federal government on scientific and technical matters. Dr. Ralph J. Cicerone is president of the National Academy of Sciences.

The **National Academy of Engineering** was established in 1964, under the charter of the National Academy of Sciences, as a parallel organization of outstanding engineers. It is autonomous in its administration and in the selection of its members, sharing with the National Academy of Sciences the responsibility for advising the federal government. The National Academy of Engineering also sponsors engineering programs aimed at meeting national needs, encourages education and research, and recognizes the superior achievements of engineers. Dr. C. D. Mote, Jr., is president of the National Academy of Engineering.

The **Institute of Medicine** was established in 1970 by the National Academy of Sciences to secure the services of eminent members of appropriate professions in the examination of policy matters pertaining to the health of the public. The Institute acts under the responsibility given to the National Academy of Sciences by its congressional charter to be an adviser to the federal government and, upon its own initiative, to identify issues of medical care, research, and education. Dr. Victor J. Dzau is president of the Institute of Medicine.

The **National Research Council** was organized by the National Academy of Sciences in 1916 to associate the broad community of science and technology with the Academy's purposes of furthering knowledge and advising the federal government. Functioning in accordance with general policies determined by the Academy, the Council has become the principal operating agency of both the National Academy of Sciences and the National Academy of Engineering in providing services to the government, the public, and the scientific and engineering communities. The Council is administered jointly by both Academies and the Institute of Medicine. Dr. Ralph J. Cicerone and Dr. C. D. Mote, Jr., are chair and vice chair, respectively, of the National Research Council.

www.national-academies.org

COMMITTEE ON THE SCIENCE OF TEAM SCIENCE

Acknowledgments

The committee and staff thank the many individuals and organizations who assisted us in our work and without whom this study could not have been completed. First, we acknowledge the generous support of the National Science Foundation and Elsevier.

Many individuals at the National Research Council (NRC) assisted the committee. We thank Patricia Morison, who provided assistance in developing a clear and concise report summary, and Kirsten Sampson-Snyder, who shepherded the report through the NRC review process. We are grateful to Mickelle Rodriguez, who arranged logistics for three committee meetings, and Tenee Davenport, who assisted with final preparation of the report.

This report has been reviewed in draft form by individuals chosen for their diverse perspectives and technical expertise, in accordance with procedures approved by the NRC's Report Review Committee. The purpose of this independent review is to provide candid and critical comments that will assist the institution in making its published report as sound as possible and to ensure that the report meets institutional standards for objectivity, evidence, and responsiveness to the study charge. The review comments and draft manuscript remain confidential to protect the integrity of the deliberative process. We thank the following individuals for their review of this report: Barry C. Barish, Department of Physics, California Institute of Technology; Edward J. Hackett, School of Human Evolution and Social Change, Arizona State University; Christine Hendren, Center for the Environmental Implications of NanoTechnology, Duke University; Nina G. Jablonski, Department of Anthropology, Pennsylvania State University; Barbara V. Jacak, Nuclear Science Division, Lawrence Berkeley National Laboratory;

Robert P. Kirshner, Department of Astronomy, Harvard-Smithsonian Center for Astrophysics; Julie T. Klein, Department of English, Wayne State University; Marshall Scott Poole, Institute for Computing in the Humanities, Arts, and Social Sciences, Department of Communication, University of Illinois Urbana-Champaign; Maritza R. Salazar, Organizational Behavior, Claremont Graduate University; Wesley M. Shrum, Department of Sociology, Louisiana State University, and Agricultural and Mechanical College; Kathryn C. Zoon, Division of Intramural Research, National Institute of Allergy and Infectious Diseases.

Although the reviewers listed above provided many constructive comments and suggestions, they were not asked to endorse the content of the report, nor did they see the final draft of the report before its release. Huda Akil, The Molecular and Behavioral Neuroscience Institute, University of Michigan, and Barbara Torrey, visiting scholar and guest researcher, Division of Behavioral and Social Research, National Institute on Aging, oversaw the review of this report. Appointed by the NRC, they were responsible for making certain that an independent examination of this report was carried out in accordance with institutional procedures and that all review comments were carefully considered. Responsibility for the final content of this report rests entirely with the authors and the institution.

Finally, we thank our colleagues on the committee for their enthusiasm, hard work, and collaborative spirit in thinking through the conceptual issues and challenges associated with addressing the charge to the study committee and in writing this report.

Nancy J. Cooke, *Chair*
Margaret L. Hilton, *Study Director*
Committee on the Science of Team Science

Contents

ix

Summary

Over the past six decades, as scientific and social challenges have become more complex and scientific knowledge and methods have advanced, scientists have increasingly joined with colleagues in collaborative research referred to as team science (see Box S-1). Today 90 percent of all science and engineering publications are authored by two or more individuals. The size of authoring teams has expanded as individual scientists, funders, and universities have sought to investigate multifaceted problems by engaging more individuals. Most articles are now written by 6 to 10 individuals from more than one institution.

Team science has led to scientific breakthroughs that would not otherwise have been possible, such as the discovery of the transistor effect, the development of antiretroviral medications to control AIDS, and confirmation of the existence of dark matter. At the same time, conducting research collaboratively can introduce challenges; for example, while the increasing size of team-based research projects brings greater scientific expertise and more advanced instrumentation to a research question, it also increases the time required for communication and coordination of work. If these challenges are not recognized and addressed, then projects may fail to achieve their scientific goals. To provide guidance in addressing these challenges, the National Science Foundation (NSF) requested that the National Research Council (NRC) appoint a committee of experts to conduct a consensus study that would "recommend opportunities to enhance the effectiveness of collaborative research in science teams, research centers, and institutes." Elsevier also provided funding for the study. The full charge to the Committee on the Science of Team Science is shown in Box S-2.

BOX S-1
Definitions

- **Team science** – Scientific collaboration, i.e., research conducted by more than one individual in an interdependent fashion, including research conducted by small teams and larger groups.
- **Science teams** – Most team science is conducted by 2 to 10 individuals, and we refer to entities of this size as science teams.
- **Larger groups** – We refer to more than 10 individuals who conduct team science as larger groups.* These larger groups are often composed of many smaller science teams, and a few of them include hundreds or even thousands of scientists. Such very large groups typically possess a differentiated division of labor and an integrated structure to coordinate the smaller science teams; entities of this type are referred to as organizations in the social sciences.
- **Team effectiveness** (also referred to as **team performance**) – A team's capacity to achieve its goals and objectives. This capacity to achieve goals and objectives leads to improved outcomes for the team members (e.g., team member satisfaction and willingness to remain together), as well as outcomes produced or influenced by the team. In a science team or larger group, the outcomes include new research findings or methods and may also include translational applications of the research.

---

*Larger groups of scientists sometimes refer to themselves as "science teams."

To create a framework for this study, the committee first defined the activity of team science and the groups that carry it out. The committee's definitions reflect prior research that has defined a "team" as two or more individuals with different roles and responsibilities, who interact socially and interdependently within an organizational system to perform tasks and accomplish common goals. Because this prior research has focused on small teams typically including 10 or fewer members, similar in size to most science teams, we refer to a group of 10 or fewer scientists as a "science team." Recognizing that what is important for successful collaboration changes dramatically as the number of participants grows, we refer to groups of more than 10 scientists as "larger groups of scientists" or simply "larger groups."

Although team science is growing rapidly, individual scientists continue to make critical contributions and important discoveries, as exemplified by Stephen Hawking's stream of new insights into the nature of the universe. Public and private funders with finite budgets must make decisions about whether to develop individual investigator or team approaches, and, if a team approach is selected, the scale and scope of the project. Similarly, individual scientists must make decisions about whether to invest time and energy in collaborative projects or to focus on individual investigations. It is important for scientists and other stakeholders to strategically consider the particular research question, subject matter, and intended scientific and/or policy goals when determining whether a team science approach is appropriate, and if so, the suitable size, duration, and structure of the project or projects.

BOX S-2
Charge to the Committee on the Science of Team Science

An ad hoc committee will conduct a consensus study on the science of team science to recommend opportunities to enhance the effectiveness of collaborative research in science teams, research centers, and institutes. The Science of Team Science is a new interdisciplinary field that empirically examines the processes by which large and small scientific teams, research centers, and institutes organize, communicate, and conduct research. It is concerned with understanding and managing circumstances that facilitate or hinder the effectiveness of collaborative research, including translational research. This includes understanding how teams connect and collaborate to achieve scientific breakthroughs that would not be attainable by either individual or simply additive efforts.

The committee will consider factors such as team dynamics, team management, and institutional structures and policies that affect large and small science teams. Among the questions the committee will explore are

1. How do individual factors (e.g., openness to divergent ideas) influence team dynamics (e.g., cohesion), and how, in turn, do both individual factors and team dynamics influence the effectiveness and productivity of science teams?
2. What factors at the team, center, or institute level (e.g., team size, team membership, geographic dispersion) influence the effectiveness of science teams?
3. How do different management approaches and leadership styles influence the effectiveness of science teams?
4. How do current tenure and promotion policies acknowledge and provide incentives to academic researchers who engage in team science?
5. What factors influence the productivity and effectiveness of research organizations that conduct and support team and collaborative science, such as research centers and institutes? How do such organizational factors as human resource policies and practices and cyber infrastructure affect team and collaborative science?
6. What types of organizational structures, policies, practices, and resources are needed to promote effective team science in academic institutions, research centers, industry, and other settings?

In order to address these questions, the committee identified, assembled, and reviewed many sources of relevant scientific research. When examining how individual- and team-level factors are related to effectiveness, the committee drew for the most part on two scientific fields that have contributed diverse methodological and conceptual approaches. Together, these fields provide cumulative empirical knowledge to assist scientists, administrators, funding agencies, and policy makers in improving the effectiveness of team science. The first is what has become known as "the science of team science," an emerging, interdisciplinary field focusing specifically on team science. The second is the large and robust body of social science research on groups and teams in contexts outside of science, such as military teams, industrial research and development teams, production and sales teams, and professional sports teams.

In reviewing the research on teams outside of science, the committee found that teams in these other contexts increasingly incorporate key features that create challenges for team science, as discussed below. This research has identified approaches to enhance team effectiveness that have been translated and extended across contexts (e.g., from aviation teams to health care teams). Therefore, based on the similarities in challenges and processes between teams in science and in other contexts and the history of generalization of team research across contexts, the committee assumes that research on teams in other contexts provides a rich foundation of knowledge that can inform strategies for improving the effectiveness of team science. The research on teams in other contexts has frequently focused on small teams, typically including 10 or fewer individuals, making it more applicable to science teams than to larger groups. However, larger groups of scientists (e.g., participants in a research center) typically are composed of multiple teams, and the research on teams in other contexts is also applicable to these teams.

When examining how organizational- and institutional-level factors are related to team effectiveness, the committee reviewed case studies of geographically distributed teams and larger groups of scientists and other professionals; the business management and leadership literatures; sociology; economics; university case studies; and science policy studies. The committee also drew on the emerging evidence from the science of team science, which focuses on not only the team level, but also the organizational, institutional, and policy levels.

Funding agencies, policy makers, scientists, and leaders of teams and larger groups all need information on how to effectively manage these projects. The first step toward increased effectiveness is to gain understanding of the factors that facilitate or hinder team science and how these factors can be leveraged to improve the management, administration, and funding of team science. Although research is emerging from the science of team science, from the research on teams, and from many other fields, this research is fragmented. Team science practitioners may have difficulty assembling, understanding, and applying the insights scattered across different research fields. This report integrates and translates the relevant research to support 13 conclusions and 9 recommendations and to identify areas requiring further research, as discussed below. Table S-1, at the end of this Summary, repeats the recommendations, specifying the individuals or organizations (e.g., team science leaders, universities) who should take action, the actions, and the desired outcomes.

KEY FEATURES THAT CREATE CHALLENGES FOR TEAM SCIENCE

Based on its review of the research evidence, information from team science practitioners, and its own expert judgment, the committee identified seven features that can create challenges for team science. Each feature represents one end of a continuous dimension. For example, large size is one end of the team or group size dimension. Science teams and larger groups often need to incorporate one or more of these features to address their particular research goals, but the features also pose challenges that are important to carefully manage. The committee returns to these seven features throughout this report in interpreting the implications of the research.

- *High diversity of membership.* Addressing complex scientific problems can require contributions from different disciplines, communities, or professions. Science teams or larger groups sometimes include community or industry stakeholders to facilitate translation of the research into practical applications (e.g., doctors or product development specialists). In addition, reflecting the changing demographics of the U.S. population and the globalization of the scientific workforce, team or group members may be diverse in age, gender, culture, religion, or ethnicity. Diverse team members may lack a common vocabulary, posing a challenge to effectively communicating about the research goals and deciding how to work together to accomplish scientific tasks.
- *Deep knowledge integration.* All science teams and larger groups integrate information to some extent as the members apply their unique knowledge and skills to the shared research problem. This challenge increases in interdisciplinary or transdisciplinary teams. *Interdisciplinary* research integrates the data, tools, perspectives, and theories of two or more disciplines to advance understanding or solve problems. *Transdisciplinary* research aims to deeply integrate and also transcend disciplinary approaches to generate

fundamentally new conceptual frameworks, theories, models, and applications. It can be difficult for the members of such teams or larger groups to share and build on each other's knowledge across the boundaries of their respective disciplines.

- *Large size.* Science and engineering teams and larger groups, as reflected in publications, have consistently expanded in size over the past 60 years. Larger size can enhance productivity by distributing the work across more individuals, but it also magnifies the burden of communicating and coordinating tasks among a larger number of individuals. Scientists participating in larger groups have fewer opportunities than those working in smaller teams to meet and work with other group members face-to-face in ways that build trust and shared understanding of project goals and the roles of other group members.

- *Goal misalignment with other teams.* Large groups of scientists, such as research centers and institutes, typically include multiple science teams engaged in research projects relevant to the higher-level research or translational goals of the center or institute. Each individual team brings valuable insights, methods, and perspectives and may have its own distinct goals. If the goals of these teams are not aligned, then this can generate conflict, requiring careful management

- *Permeable boundaries.* The boundaries of science teams and larger groups are often permeable, reflecting changes in the project goals over time. The membership of a group or team may change as the project moves from one phase, requiring a certain type of expertise, to another that may require different expertise. Although these changes have the benefit of matching expertise to scientific or translational problems as they arise, they can also create challenges for effective team or group interaction.

- *Geographic dispersion.* Most science teams and larger groups are geographically dispersed, with members located across multiple universities or research institutions. Although crossing institutional boundaries can bring needed expertise, scientific instrumentation, datasets, or other valuable resources to a science team or larger group, it also requires greater reliance on electronic modes of communication, with attendant challenges. In addition, the team or larger group may find it difficult to coordinate work across institutions with varying work styles, time zones, and cultural expectations about scientific work.

- *High task interdependence.* One of the defining features of a team is that the members are dependent on each other to accomplish a shared task. All team science projects aim to tap the benefits of interdependent, collaborative research, yet designing and conducting interdependent tasks that draw on and integrate the unique talents of the individual team or larger group members to accomplish shared goals can be challenging. Greater task interdependence among team or group members can lead to more opportunities for conflict, and when geographically dispersed members must perform highly interdependent tasks, greater coordination and communication efforts may be required.

Each science team or larger group is unique in the extent to which it is characterized by one or more of these features. As a given team or group incorporates more of these key features—for instance, high diversity of membership and geographic dispersion—so do the accompanying challenges and the attendant need to understand and carefully manage them. As noted above, it is important to strategically consider the particular research question, subject matter, and intended goals when determining the approach, suitable size, and other features of a research project.

IMPROVING TEAM AND GROUP EFFECTIVENESS

Research on teams in non-science contexts has identified strategies for improving effectiveness that can be translated and applied to help science teams and larger groups navigate the challenges involved in team science.

CONCLUSION. *A strong body of research conducted over several decades has demonstrated that team processes (e.g., shared understanding of team goals and member roles, conflict) are related to team effectiveness. Actions and interventions that foster positive team processes offer the most promising route to enhance team effectiveness; they target three aspects of a team: team composition (assembling the right individuals), team professional development, and team leadership.*

Team Composition

Assembling and composing the team provides the raw building material for an effective team and therefore is a critical step requiring careful management, but it is only the first step.

CONCLUSION. *Research to date in non-science contexts has found that team composition influences team effectiveness, and this relationship depends on the complexity of the task, the degree of interdependence among team members, and how long the team is together. Task-relevant diversity is critical and has a positive influence on team effectiveness.*

CONCLUSION. *Task analytic methods developed in non-science contexts and research networking tools developed in science contexts allow practitioners to consider team composition systematically.*

RECOMMENDATION 1: Team science leaders and others involved in assembling science teams and larger groups should consider making use of task analytic methods (e.g., task analysis, cognitive modeling, job analysis, cognitive work analysis) and tools that help identify the knowledge, skills, and attitudes required for effective performance of the project so that task-related diversity among team or group members can best match project needs. They should also consider applying tools such as research networking systems designed to facilitate assembly of science teams and partner with researchers to evaluate and refine these tools and task analytic methods.

Team Professional Development

Once a science team or larger group has been assembled, it faces the challenge of integrating the members' knowledge to achieve its scientific goals. Knowledge integration, along with shared understanding of research goals and member roles, can be facilitated by formal professional development programs (referred to in the research literature as training programs).

CONCLUSION. *Research in contexts outside of science has demonstrated that several types of team professional development interventions (e.g., knowledge development training to increase sharing of individual knowledge and improve problem solving) improve team processes and outcomes.*

RECOMMENDATION 2: Team-training researchers, universities, and science team leaders should partner to translate, extend, and evaluate the promising training strategies, shown to improve the effectiveness of teams in other contexts, to create professional development opportunities for science teams.

Although research has demonstrated that training for current team members can increase team effectiveness, educational programs designed to prepare students for future team science have only recently emerged and have not yet been systematically evaluated.

CONCLUSION. *Colleges and universities are developing cross-disciplinary programs designed to prepare students for team science, but little empirical research is available on the extent to which participants in such programs develop the competencies they target. Research to date has not shown whether the acquisition of the targeted competencies contributes to team science effectiveness.*

Leadership for Team Science

Currently, most leaders of science teams and larger groups are appointed to their positions based solely on scientific expertise and lack formal leadership training. At the same time, an extensive body of research on organizational and team leadership has illuminated leadership styles and behaviors that foster positive interpersonal processes, thereby enhancing effectiveness in teams and larger groups. These effective leadership styles and behaviors can be acquired.

CONCLUSION. *Fifty years of research on team and organizational leadership in contexts other than science provide a robust foundation of evidence to guide professional development for leaders of science teams and larger groups.*

RECOMMENDATION 3: Leadership researchers, universities, and leaders of team science projects should partner to translate and extend the leadership literature to create and evaluate science leadership development opportunities for team science leaders and funding agency program officers.

Supporting Virtual Collaboration

As science attempts to answer bigger and bigger questions, it is increasingly likely that the people participating in research projects reside in different locations, institutions, and even countries. This geographic dispersion can lead to challenges, particularly with communication and coordination. Addressing the special challenges such groups and teams encounter requires effective leadership and technology.

CONCLUSION. *Research on geographically dispersed teams and larger groups of scientists and other professionals has found that communicating progress, obstacles, and open issues and developing trust are more challenging relative to face-to-face teams and larger groups. These limitations of virtual collaboration may not be obvious to members and leaders of the team or group.*

RECOMMENDATION 4: Leaders of geographically dispersed science teams and larger groups should provide activities shown by research to help all participants develop shared knowledge (e.g., a common vocabulary and work style). These activities should include team professional development opportunities that promote knowledge sharing (see Recommendation #2 above). Leaders should also consider the feasibility of assigning some tasks to semi-independent units at each location to reduce the burden of constant electronic communication.

CONCLUSION. *Technology for virtual collaboration often is designed without a true understanding of users' needs and limitations and even when a suite of appropriate technologies is available, users often do not recognize and use its full capabilities. These related problems may thus impede such collaboration.*

RECOMMENDATION 5: When selecting technologies to support virtual science teams or larger groups, leaders should carefully evaluate the needs of the project, and the ability of the individual participants to embrace new technologies. Organizations should promote human-centered collaboration technologies, provide technical staff, and encourage use of the technologies by providing ongoing training and technology support.

Organizational Supports for Team Science

Science teams and larger groups are often housed within universities. In these complex organizations, faculty members' decisions about whether and when to participate in team science are influenced by various contexts and cultures including the department, the college, the institution as a whole, and external groups, such as disciplinary societies. Formal rewards and incentive structures, reflecting these various cultures, currently tend to focus on individual research contributions. Some universities have recently sought to promote interdisciplinary team science by, for example, merging disciplinary departments to create interdisciplinary research centers or schools, providing seed grants, and forging partnerships with industry. However, little is known about the impact of these efforts, while the lack of recognition and rewards for team science can deter faculty members from joining science teams or larger groups.

CONCLUSION. *Various research universities have undertaken new efforts to promote interdisciplinary team science, such as merging disciplinary departments to create interdisciplinary research centers or schools. However, the impact of these initiatives on the amount and quality of team science research remains to be systematically evaluated.*

CONCLUSION. *University policies for promotion and tenure review typically do not provide comprehensive, clearly articulated criteria for evaluating individual contributions to team-based research. The extent to which researchers are rewarded for team-based research varies widely across and within universities. Where team-based research is not rewarded, young faculty may be discouraged from joining those projects.*

In a few isolated cases, universities have developed new policies for assessing individual contributions to team science. At the same time, research has begun to characterize the various types of individual contributions and develop software systems that would identify each individual's role during the process of submitting and publishing an article. This work can inform new efforts by universities and disciplinary associations.

RECOMMENDATION 6: **Universities and disciplinary associations should proactively develop and evaluate broad principles and more specific criteria for allocating credit for team-based work to assist promotion and tenure committees in reviewing candidates.**

Funding for Team Science

CONCLUSION. *Public and private funders are in the position to foster a culture within the scientific community that supports those who want to undertake team science, not only through funding, but also through white papers, training workshops, and other approaches.*

RECOMMENDATION 7: **Funders should work with the scientific community to encourage the development and implementation of new collaborative models, such as research networks and consortia; new team science incentives, such as academic rewards for team-based research (see Recommendation #6); and resources (e.g., online repositories of information on improving the effectiveness of team science and training modules).**

CONCLUSION. *Funding agencies are inconsistent in balancing their focus on scientific merit with their consideration of how teams and larger groups are going to execute the work (collaborative merit). The*

Funding Opportunity Announcements they use to solicit team science
proposals often include vague language about the type of collaboration
and the level of knowledge integration they seek in proposed research.

Currently, proposals for team science research grants do not address
how the participating scientists will collaborate. Research has shown that
engaging team members in explicit discussions of how to coordinate and
integrate their work enhances effectiveness, as does the development of
team charters that outline team directions, roles, and processes. In addition,
research has found that large, multi-institutional groups of scientists often
benefit from establishing formal contracts outlining roles and assignments.
Collaboration plans build on both team charter and contract concepts,
promising to enhance the effectiveness of team science.

> RECOMMENDATION 8: Funders should require proposals for team-based
> research to present collaboration plans and provide guidance to sci-
> entists for the inclusion of these plans in their proposals, as well as
> guidance and criteria for reviewers' evaluation of these plans. Funders
> should also require authors of proposals for interdisciplinary or trans-
> disciplinary research projects to specify how they will integrate disci-
> plinary perspectives and methods throughout the life of the research
> project.

ADVANCING RESEARCH ON THE
EFFECTIVENESS OF TEAM SCIENCE

The committee's review of the research related to the study charge
identified several areas in which further research is needed to enhance understanding of team science and improve its effectiveness.

Continued research and evaluation will be needed to refine and enhance the actions, interventions, and policies recommended in this report.
At the same time, research is needed to enhance basic understanding of
team science processes as the foundation for developing new interventions.
Funders of scientific research, policy makers, and the scientific community
need appropriate criteria for evaluating the potential (ex-ante) and achieved
(ex-post) outcomes of team science. In addition, funders and policy makers
would benefit from more rigorous evaluations incorporating experimental
or quasi-experimental methods to generate stronger evidence that team-
based research approaches increase research productivity beyond what
would have been accomplished by the individual scientists working alone
or as members of a different team or group. An essential first step toward
meeting these goals is to increase researchers' *access* to practicing scientists
to study their interactions and innovations. In sum, advancing the research

on the effectiveness of science teams and larger groups will require funding, as well as the dedication of research organizations, team science leaders, and the scientific community as a whole.

> CONCLUSION. *Targeted research is needed to evaluate and refine the tools, interventions, and policies recommended above, along with more basic research, to guide continued improvement in the effectiveness of team science. However, few if any funding programs support research on the effectiveness of science teams and larger groups.*

> RECOMMENDATION 9: **Public and private funders should support research on team science effectiveness through funding. As critical first steps, they should support ongoing evaluation and refinement of the interventions and policies recommended above and research on the role of scientific organizations (e.g., research centers, networks) in supporting science teams and larger groups. They should also collaborate with universities and the scientific community to facilitate researchers' access to key team science personnel and datasets.**

Promising new research methods and approaches can be applied to implement this recommendation. Complex adaptive systems theory offers a route to understand how behaviors, actions, and reactions at one level of a team science system (e.g., the individual level) affect actions at other system levels (e.g., the team level) and the emergent behavior of the system as a whole. To study team and group dynamics, members can be equipped with small electronic sensor badges that record data on their interactions. Similarly, electronic communication data, such as emails and texts, can be recorded and analyzed. These new forms of data can be creatively combined with publication data to examine the relationship between team or group processes and outcomes. Such approaches will facilitate further research to deepen understanding of team science and enhance its effectiveness.

TABLE S-1 Recommended Actions and Desired Outcomes

Actor	Recommended Action	Desired Outcome
Leaders of Science Teams and Groups	• **Recommendation 1:** Consider applying analytic methods and tools to guide team composition and assembly.	• Match mix of participants to project needs to enhance scientific/translational effectiveness.
	• **Recommendation 2:** Partner with team-training researchers and universities to create and evaluate professional development opportunities for science teams.	• Foster positive team processes and thereby enhance effectiveness.
	• **Recommendation 3:** Partner with leadership researchers and universities to create and evaluate science leadership development opportunities.	• Increase capacity of team and group leaders and funding agency staff to facilitate positive team processes and thereby enhance effectiveness.
Leaders of Geographically Dispersed Science Teams and Larger Groups	• **Recommendation 4:** Provide activities to develop shared knowledge among all participants, including team professional development opportunities. o Consider the feasibility of dividing up some of the work.	• Develop shared vocabularies and work routines across locations to enhance effectiveness. Foster knowledge sharing and knowledge integration.
	• **Recommendation 5:** Select collaboration technologies based on careful evaluation of their readiness, project needs, and team members' ability to use them. Access technology training and support.	• Reduce burden of constant electronic communication to allow participant to focus on scientific tasks. • Optimize use of the most appropriate collaboration technologies to enhance effectiveness.

TABLE S-1 Continued

Actor	Recommended Action	Desired Outcome
Universities and Other Scientific Organizations	• **Recommendation 2:** Partner with team-training researchers and universities to create and evaluate professional development opportunities for science teams.	• Foster positive team processes and thereby enhance effectiveness.
	• **Recommendation 3:** Partner with leadership researchers and team science leaders to create and evaluate leadership development opportunities.	• Increase capacity of team and group leaders and funding agency program officers to facilitate positive team processes and thereby enhance effectiveness.
	• **Recommendation 6:** Collaborate with disciplinary associations to develop broad principles and more specific criteria for allocating credit for team-based work; collaborate with researchers to evaluate the role of such principles.	• Remove a barrier that discourages young faculty from participating in team science.
Public and Private Funders	• **Recommendation 7:** Work with the scientific community to encourage new collaborative models, remove disincentives to participate in team science, and provide informational resources.	• Foster culture change in the scientific community to reduce barriers to team science.
	• **Recommendation 8:** Require authors of proposals for team-based research to include collaboration plans and, for interdisciplinary or transdisciplinary projects, specify how they will foster deep knowledge integration over the life of the research project.	• Encourage project leaders to plan not only for the scientific/ technical aspects of the research but also for the collaborative/ interpersonal aspects.

Continued

TABLE S-1 Continued

Actor	Recommended Action	Desired Outcome
	• **Recommendation 9:** Support further research on team science effectiveness and facilitate researchers' access to key personnel and data.	• Facilitate evaluation and improvement of the tools, actions, and interventions recommended above as well as more "basic" research to enhance team science effectiveness and speed scientific discovery.
Researchers	• **Recommendation 1:** Partner with team science leaders to evaluate and improve analytic methods and tools for team assembly.	• Improve methods and tools to match mix of participants with project needs to enhance scientific/translational effectiveness.
	• **Recommendation 2:** Partner with science team leaders and universities to create and evaluate professional development opportunities for science teams.	• Foster positive team processes and thereby enhance effectiveness.
	• **Recommendation 3:** Partner with team science leaders and universities to create and evaluate team science leadership development opportunities.	• Increase capacity of team and group leaders and funding agency staff to facilitate positive team processes and thereby enhance effectiveness.
	• **Recommendation 6:** Collaborate with universities and disciplinary associations to evaluate the role of new principles and criteria for allocating credit for team science in reducing barriers to participation in team science.	• Remove a barrier that discourages young faculty who are interested in team science from joining teams or larger groups.
Scientific Community	• **Recommendation 6:** Collaborate with universities to develop and evaluate broad principles and more specific criteria for allocating credit for team-based work.	• Remove a barrier that discourages young faculty who are interested in team science from joining teams or larger groups.
	• **Recommendation 7:** Work with public and private funders to encourage new collaborative models, remove disincentives to team science, and access resources.	• Foster culture change in the scientific community to reduce barriers to team science.

Part I

Setting the Stage

1

Introduction

The past half-century has witnessed a dramatic increase in the scale and complexity of scientific research that has yielded exciting discoveries about natural phenomena and an array of practical applications, improving human health and the quality of life while fueling the growth of dynamic industries, such as pharmaceuticals, biotechnology, personal computing, advanced manufacturing, and software development.

The growing scale of science has been accompanied by a dramatic shift toward collaborative research referred to as "team science" defined further below. Studying the corpus of 19.9 million research articles across the fields of science and engineering, social science, and arts and humanities (Web of Science) and 2.1 million patent records (National Bureau of Economic Research) for more than five decades, Wuchty, Jones, and Uzzi (2007) discovered that the propensity for teamwork is greatest in the life and physical sciences but is also rapidly increasing in the social sciences. The authors found that 80 percent of all science and engineering publications were written by teams of two or more authors in 2000. The Committee on the Science of Team Science updated the database and trend analysis to find that the share of all papers written by two or more authors increased to 90 percent by the year 2013 (see Figure 1-1).

Wuchty, Jones, and Uzzi (2007) also found that the size of science and engineering authoring teams consistently expanded over the period, from a mean of less than 2 members in 1960 to 3.5 members in 2000. In a follow-up study, Jones, Wuchty, and Uzzi (2008) found that the rapid growth in team-based publications was due to an increase in publications

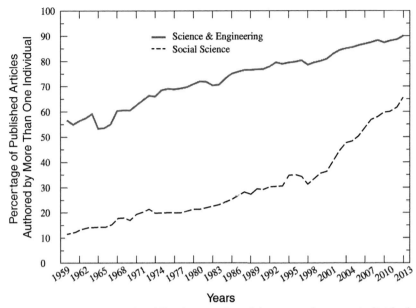

FIGURE 1-1 Percentage of publications authored by more than one individual, 1960–2013.

by authors from multiple institutions, showing that team-based research increasingly crosses institutional and geographic boundaries.

PURPOSE OF THIS REPORT

Although team science is growing rapidly, it can be more challenging than solo science. For example, the increasing size of research teams and groups (Wuchty, Jones, and Uzzi, 2007) brings greater scientific expertise and more advanced instrumentation to a research question but also increases the amount of time required for communication and coordination of work among a greater number of individuals (see further discussion below). Given the growth of team science, evidence-based guidance is needed for addressing the challenges associated with these approaches and achieving their potential to more rapidly solve scientific and societal problems. To provide such guidance, the National Science Foundation (NSF) requested the National Research Council (NRC) to convene an expert committee and address the charge presented in Box 1-1. The study is also supported by Elsevier.

To address this charge, the committee identified, assembled, and re-

---

BOX 1-1
Charge to the Committee on the Science of Team Science

An ad hoc committee will conduct a consensus study on the science of team science to recommend opportunities to enhance the effectiveness of collaborative research in science teams, research centers, and institutes. The Science of Team Science is a new interdisciplinary field that empirically examines the processes by which large and small scientific teams, research centers, and institutes organize, communicate, and conduct research. It is concerned with understanding and managing circumstances that facilitate or hinder the effectiveness of collaborative research, including translational research. This includes understanding how teams connect and collaborate to achieve scientific breakthroughs that would not be attainable by either individual or simply additive efforts.

The committee will consider factors such as team dynamics, team management, and institutional structures and policies that affect large and small science teams. Among the questions the committee will explore are

1. How do individual factors (e.g., openness to divergent ideas) influence team dynamics (e.g., cohesion), and how, in turn, do both individual factors and team dynamics influence the effectiveness and productivity of science teams?
2. What factors at the team, center, or institute level (e.g., team size, team membership, geographic dispersion) influence the effectiveness of science teams?
3. How do different management approaches and leadership styles influence the effectiveness of science teams?
4. How do current tenure and promotion policies acknowledge and provide incentives to academic researchers who engage in team science?
5. What factors influence the productivity and effectiveness of research organizations that conduct and support team and collaborative science, such as research centers and institutes? How do such organizational factors as human resource policies and practices and cyber infrastructure affect team and collaborative science?
6. What types of organizational structures, policies, practices, and resources are needed to promote effective team science in academic institutions, research centers, industry, and other settings?

---

viewed many sources of relevant scientific research. When focusing on individual- and team-level factors, the committee drew primarily on the robust evidence on teams in contexts outside of science, supplemented by the emerging evidence from the new interdisciplinary field of the science of team science. When focusing on organizational- and institutional-level factors, it drew on leadership literature, case studies of geographically distributed teams and larger groups of scientists and other professionals, business management literature, sociology, economics, and science policy studies. The committee's analysis of organizational and institutional factors was also supplemented by the emerging evidence from the science of team science, which focuses not only on the team level, but also on the organizational, institutional, and policy levels. This report is the culmination of an intensive study conducted to determine what is currently known about the processes and products of team science, and the circumstances under which investments in team-based research are most likely to yield intellectually novel discoveries and demonstrable improvements in contemporary social, environmental, and public health problems.

DEFINING KEY TERMS

To create a framework for this study, the committee first defined the activity of team science and the groups that carry it out (see Box 1-2). The committee's definitions reflect prior research that has defined a "team" as two or more individuals with different roles and responsibilities, who interact socially and interdependently within an organizational system to perform tasks and accomplish common goals. Because this prior research

---

BOX 1-2
Definitions

- **Team science** – Scientific collaboration, i.e., research conducted by more than one individual in an interdependent fashion, including research conducted by small teams and larger groups.
- **Science teams** – Most team science is conducted by 2 to 10 individuals, and we refer to entities of this size as science teams.
- **Larger groups** – We refer to more than 10 individuals who conduct team science as larger groups.* These larger groups are often composed of many smaller science teams, and a few of them include hundreds or even thousands of scientists. Such very large groups typically possess a differentiated division of labor and an integrated structure to coordinate the smaller science teams; entities of this type are referred to as organizations in the social sciences.
- **Team effectiveness** (also referred to as **team performance**) – A team's capacity to achieve its goals and objectives. This capacity to achieve goals and objectives leads to improved outcomes for the team members (e.g., team member satisfaction and willingness to remain together), as well as outcomes produced or influenced by the team. In a science team or larger group, the outcomes include new research findings or methods and may also include translational applications of the research.

---

*Larger groups of scientists sometimes refer to themselves as "science teams."

has focused on small teams typically including 10 or fewer members, similar in size to most science teams, we refer to a group of 10 or fewer scientists as a "science team." Recognizing that what is important for successful collaboration changes dramatically as the number of participants grows, we refer to groups of more than 10 scientists as "larger groups of scientists" or simply "larger groups."

Although an individual investigator can master and integrate knowledge from diverse disciplines—for example, physicist Albert Einstein used mathematics, specifically Riemann geometry to create his new General Theory of Relativity—this process has become more difficult over the past four decades, because of the rapid growth of specialized knowledge in all fields of science and engineering (Jones, 2009). A scientist interested in investigating questions that require knowledge beyond her or his narrow specialization may prefer to team up with colleagues to obtain complementary expertise, rather than spending years mastering another discipline.

Science teams and larger groups vary in the extent to which they include or integrate the knowledge of experts from different disciplines or professions to achieve their scientific and, when relevant, translational goals. These varying degrees of integration have been classified as *unidisciplinary*, *multidisciplinary*, *interdisciplinary*, and *transdisciplinary* research approaches (see Figure 1-2). *Unidisciplinary research* relies on the methods, concepts, and approaches of a single discipline. In *multidisciplinary research*, each discipline makes separate contributions in an additive way. *Interdisciplinary research* integrates "information, data, techniques, tools, perspectives, concepts, and/or theories from two or more disciplines . . . to advance fundamental understanding or to solve problems" (National Academy of Sciences, National Academy of Engineering, and Institute of Medicine, 2005, p. 26). Interdisciplinary research has grown over the past three decades (Frickel and Jacobs, 2009; Porter and Rafels, 2009), reflecting the need for multiple disciplinary perspectives to address complex scientific and societal problems. *Transdisciplinary research* integrates but also transcends disciplinary approaches, as follows (Stokols, Hall, and Vogel, 2013, p. 5):

> [T]he TD [transdisciplinary] approach entails not only the integration of approaches but also the creation of fundamentally new conceptual frameworks, hypotheses, and research strategies that synthesize diverse approaches and ultimately extend beyond them to transcend preexisting disciplinary boundaries.

Some, but not all transdisciplinary research projects emphasize translation of research findings into practical solutions to social problems and include societal stakeholders (e.g., health professionals, business representatives) to facilitate this translation.

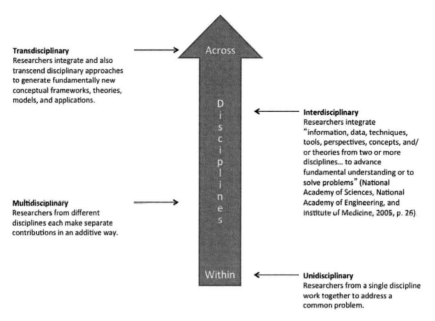

Transdisciplinary
Researchers integrate and also transcend disciplinary approaches to generate fundamentally new conceptual frameworks, theories, models, and applications.

Interdisciplinary
Researchers integrate "information, data, techniques, tools, perspectives, concepts, and/or theories from two or more disciplines... to advance fundamental understanding or to solve problems" (National Academy of Sciences, National Academy of Engineering, and Institute of Medicine, 2005, p. 26)

Multidisciplinary
Researchers from different disciplines each make separate contributions in an additive way.

Unidisciplinary
Researchers from a single discipline work together to address a common problem.

FIGURE 1-2 Levels of cross-disciplinary integration.

Since the 1980s, some parts of the scientific community have increased their use of transdisciplinary research approaches as a promising way to gain new scientific and technical insights on complex phenomena and speed application of these insights. For example, "convergence" integrates expertise from the life, physical, computational, and other sciences within a network of academic, industry, clinical, and funding partnerships to address scientific and societal challenges (National Research Council, 2014). In another example, the field of transdisciplinary sustainability studies brings together environmental scientists, policy makers, citizens, and industry representatives to frame and address multifaceted environmental challenges (Huutoniemi and Tapio, 2014). To illustrate these varying approaches to disciplinary integration, Box 1-3 provides examples from archaeology.

KEY FEATURES THAT CREATE CHALLENGES
FOR TEAM SCIENCE

Based on its review of the research evidence, information from team science practitioners, and its own expert judgment, the committee identified seven features that create challenges for team science. A given team or group may need to incorporate one or more of these features to address

BOX 1-3
Changing Research Approaches in Archaeology Teams

Much of the early history of American archaeology is characterized by unidisciplinary research. A classic example of this can be found in *Ancient Monuments of the Mississippi Valley* by Squier and Davis (1848), the first major scientific publication of the new Smithsonian Institution. In the 20th century, the important *The Fort Ancient Aspect* report by Griffin (1943), is another key example of unidisciplinary research. However, much of the research in the 20th century, especially in the second half, features multidisciplinary studies, with the nonarchaeological work often published as appendixes or separate chapters in the final publication or as separate reports. For example, the research at the ancient Maya site of Seibal, Guatemala (see Willey et al. [1975] for an introduction) included specialized scientific studies of plaster, animal bone, ceramics, and stone tools. Neutron activation analyses of ceramics undertaken at Brookhaven National Laboratory provided significant data on the sources of clays, contributing to understanding of ancient Maya economics and politics.

Interdisciplinary research in American archaeology fully emerged after World War II. An example can be found in the research on the Early Classic Period at the ancient Maya site of Copan, Honduras, that focused on the rise of the ruling dynasty of Copan. This research fully integrated diverse disciplines or approaches, such as archaeology, iconography, epigraphy, studies of human skeletal remains, bone chemistry studies, and neutron activation analyses of ceramics, among others (see Bell, Canuto, and Sharer, 2003).

To date, truly transdisciplinary studies are rare in world archaeology. One example that can be noted began with research in the Americas but has since become widespread: beginning with Lewis Binford's ethno-archaeological research among the Nunamiut peoples of Alaska, and the melding of understandings from disciplines, such as archaeology, ethnography, biology, ecology, geography, and statistics, Binford, his students, and archaeological colleagues came up with a new approach. Their transdisciplinary work yielded new insights into the nature of modern and archaic hunter-gatherer activities and settlement systems through time and space that transcended interdisciplinary research (see Binford 1978, 1980, 2001; Kelly 1995, among many others).

its particular research goals, but the features also pose challenges that are important to carefully manage. They include (1) high diversity of membership; (2) deep knowledge integration; (3) large size; (4) goal misalignment with other teams; (5) permeable team and group boundaries; (6) geographic dispersion; and (7) high task interdependence.

These features are based on levels or degrees within the team science dimensions shown in Table 1-1. The dimensions reflect variations in composition, size, and other facets of team science and do not necessarily introduce significant challenges for a team science project. However, we

TABLE 1-1 Dimensions of Team Science

Dimension	Range	
Diversity of Team or Group Membership	Homogeneous	Heterogeneous
Disciplinary Integration	Unidisciplinary	Transdisciplinary
Team or Group Size	Small (2)	Mega (1000s)
Goal Alignment Across Teams	Aligned	Divergent or misaligned
Permeable Team and Organizational Boundaries	Stable	Fluid
Proximity of Team or Group Members	Co-located	Globally distributed
Task Interdependence	Low	High

characterize certain levels or degrees along each dimension (e.g., large size) as key features that create challenges for team science, increasing the need for strategies to mitigate such challenges.

Although each team science project is unique in the extent to which it is characterized by these features, as a given project incorporates more features—for instance, the inclusion of more disciplines or large size—so do the accompanying challenges, and the imperative for better understanding how the interacting features influence research processes and outcomes, to enhance the success of the project. Science teams and larger groups are increasingly likely to incorporate one or more of these seven features because they are needed to address complex scientific and societal problems. For example, greater diversity of membership may be needed to answer particularly complex scientific questions or a large group of scientists may be needed to maximize the benefits of an investment in large instrumentation. However, these features may not always be necessary. Therefore, scientists and funders engaged in designing team science projects need not necessarily include highly diverse members or large numbers of participants (Vermeulen et al., 2010), as the costs may outweigh the benefits (Cummings et al., 2013). Rather, strategically considering the nature of the scientific problem, the readiness of the science, and other factors are important to determine the best approach and scale of a research activity.

Next, we discuss each of the seven features in more detail, with an example and more general discussion. The examples are summarized in Table 1-2.

TABLE 1-2 Key Features, Goals, and Potential Challenges of Team Science

Feature That Creates Challenges	Example Project	Project Goals Requiring Feature	Potential Challenges
High Diversity of Membership	"Social Environment, Stress, and Health" project	Reduce breast cancer by understanding and addressing its relationships with neighborhood and community factors and behavioral and biological responses.	Identify community partners and establishes positive relationships with them. Foster effective communication and coordination of tasks among individuals from different scientific disciplines and communities with their own languages and cultures.
Deep Knowledge Integration	National Institutes of Health Trans-disciplinary Research in Energetics and Cancer Centers	Understand the relationships among obesity, nutrition, physical activity, and cancer.	Require more time and effort than other research approaches. Integrate knowledge across social, behavioral, and biological disciplines with different values, terminology, methods, traditions, and work styles (Vogel et al., 2014).
Large Size	Manhattan Project to develop the atomic bomb during World War II	Aid the war effort by translating theoretical knowledge of atomic fission into a powerful weapon.	Coordinate the work of 130,000 individuals at different locations. Foster effective communication among physicists, engineers, construction workers, nuclear facility production workers, and clerical staff.

Continued

TABLE 1-2 Continued

Feature That Creates Challenges	Example Project	Project Goals Requiring Feature	Potential Challenges
Goal Misalignment with Other Teams	James Webb Space Telescope	Create the next Great Observatory to replace the Hubble Space Telescope.	Fund, manage, and align multiple academic and industry teams (James Webb Space Telescope Independent Comprehensive Review Panel, 2010; U.S. Government Accountability Office, 2012)
Permeable Team and Group Boundaries	International Maize and Wheat Improvement Center in Mexico (Cash et al., 2003)	Improve nutrition in rural Mexico and Central America by translating findings from plant science to the field.	Engage indigenous farmers in the project while also ensuring scientific rigor in the plant science research. Gain understanding of the kinds of information the farmers need so that scientific findings can be tailored to meet their needs.
Geographic Dispersion	Thirty Meter Telescope, being developed by a partnership of research institutions in the U.S., India, China, Japan, and Canada	Plan and design a powerful optic telescope enabling astronomers to study the very edge of the observable universe.	Build cohesion among experts who rarely meet face-to-face and rely heavily on electronic communication. Develop shared understanding of project goals and individual roles among scientists from nations and research institutions with different cultures, work routines, and politics.

TABLE 1-2 Continued

Feature That Creates Challenges	Example Project	Project Goals Requiring Feature	Potential Challenges
High Task Interdependence	Search for the Higgs Boson at the Large Hadron Collider in Geneva, Switzerland (see Box 6-1)	Increase understanding of subatomic particles by replicating conditions at the time of the "Big Bang."	Foster a shared appreciation of the importance of two types of highly interdependent tasks: "service" work (managing the collider, detector, global computer network etc.) and "physics" work (analysis of data leading to publications). Reach agreement among groups and individuals over new research approaches (e.g., modifications to detectors or data analysis methods).

High Diversity of Membership

The members of a science team or group may come from different disciplines, research institutions, or nations. When relevant, the members may include community or industry stakeholders (e.g., doctors or product development specialists) to facilitate the research and/or its translation into practical applications. The members may be diverse in age, gender, culture, and other demographic characteristics. For example, the Social Environment, Stress, and Health project supported by the National Institutes of Health used a community-based participatory research approach to understand relationships among neighborhood and community factors, behavioral and biological responses, and breast cancer among women living on Chicago's South Side (Hall et al., 2012a). The investigators, including natural and social scientists, conducted focus groups to learn about the beliefs, attitudes, and concerns of community members regarding breast cancer. Focus group members who were particularly committed to the research were invited to form a community advisory board as an active partner in the project. The newly evolved group, including scientists and stakeholders, worked with the community to share the research findings and identify and rank translational "action steps" to address them. Developing messages about wellness for 12- to 16-year-olds on the South Side was ranked as the most important action, a translational focus that would not have occurred to the investigators working by themselves.

A key assumption underlying the formation of interdisciplinary and transdisciplinary team science projects is that the inclusion of individuals with diverse knowledge, perspectives, and research methods will lead to scientific or translational breakthroughs that might not be achieved by a more homogenous group of individuals (e.g., National Academy of Sciences, National Academy of Engineering, and Institute of Medicine, 2005; Fiore, 2008). Research on work groups and teams provides some support for this assumption, suggesting that including individuals with diverse knowledge, expertise, and experience can increase group creativity and effectiveness but only if group members draw on each other's diverse expertise (Ancona and Caldwell, 1992; Stasser, Stewart, and Wittenbaum, 1995; Homan et al., 2008). However, encouraging members to draw on each other's diverse expertise can be challenging.

Diversity in membership—whether in terms of expertise or demographic factors—influences the group's effectiveness through its impact on group processes, such as decision making and conflict management (Bezrukova, 2013). Hence, greater diversity of membership increases the challenges facing a group by influencing these processes. High levels of diversity bring benefits, but differences among members can weaken identification with the group (Cummings et al., 2013). Members may differ in their values and motivations, shaped by their unique areas of expertise, organizational contexts, or life experiences. For example, when universities form research partnerships with private companies, the academic scientists who are rewarded for publications may have very different motivations than the industry scientists, who are rewarded for achieving specific business benchmarks (Bozeman and Boardman, 2013).

In highly diverse team science projects, communication problems can occur because of members' use of technical or scientific language that is unique to their area of expertise and therefore unfamiliar to other members. The unique languages of the disciplines reflect deeper differences in underlying assumptions, epistemologies (ways of knowing), philosophies, and approaches to science and societal problems (Eigenbrode et al., 2007). For example, laboratories in molecular biology and those in high-energy physics have very different "epistemic cultures"—the practices and beliefs that constitute each discipline's attitude toward knowledge and its way of justifying knowledge claims (Knorr-Cetina, 1999). When teams or groups fail to identify, discuss, and clarify these differences among their members, confusion and conflict can arise.

Chapter 3 highlights empirical evidence related to the team processes that underlie these challenges, which emerge from increasing diversity. Chapters 4, 5, and 6 introduce strategies for addressing them.

Deep Knowledge Integration

Knowledge integration occurs in some form within all scientific collaborations, as team or group members apply their unique knowledge and skills to the shared research problem. The process of knowledge integration can be challenging, and this challenge increases when scientific and societal questions require not only the combination, but also the deep integration of a broad set of disciplinary and, when relevant, stakeholder perspectives. Such deeper integration is fostered by interdisciplinary and transdisciplinary research approaches (Misra et al., 2011b; Salazar et al., 2012). For example, the National Cancer Institute's Transdisciplinary Research on Energetics and Cancer (TREC) initiative to integrate social, behavioral, and biological sciences to address obesity and overweight, physical inactivity, and poor diet with the goal of preventing and controlling cancer. The integrative approach led to many novel discoveries; for example, one study found that participation in a 12-month exercise program decreased oxidative stress, which is closely linked to inflammation and cancer (Vogel et al., 2014).

To achieve the goals of interdisciplinary and transdisciplinary research, it is essential to understand and address the challenges associated with the deeper levels of disciplinary integration they entail. These challenges can emerge in efforts to integrate the knowledge of members from different disciplines with different cultures, languages, and research practices (Knorr-Cetina, 1999). Participating scientists may feel uncomfortable crossing the boundaries of their home disciplines—both the physical boundaries of their disciplinary department, laboratory, or office, and the cultural boundaries that guide and focus their research activities (Klein, 2010). Not all collaborators will be ready or willing to engage in the same level of integrative work. As the degree of integration increases, individuals may face challenges with feeling the loss of disciplinary "identity" or fear of becoming a "generalist" (Salazar et al., 2012). In molecular biology, for example, scientists' identities are closely linked with the materials, techniques, instruments, and enabling theories of their research groups or laboratories, which Hackett (2005) refers to as "ensembles of technologies." These challenges can be instigated and perpetuated by organizational cultures and incentive systems (e.g., promotion and tenure policies) that reward work within a single laboratory or a single discipline (Fiore, 2008; Stokols, Hall, and Vogel, 2013).

Strategies to address these challenges and foster successful knowledge integration in science teams and larger groups are discussed in Chapters 4 through 9 of this report.

Large Size

Science and engineering teams and larger groups, as reflected in publications, have consistently expanded in size over the past five decades (Adams et al., 2005; Baker, Day, and Salas, 2006; Wuchty, Jones, and Uzzi, 2007). This trend is illustrated in Figure 1-3, which shows the frequency of papers published in each year by single authors and groups of various sizes from 1960 to 2013, based on authorship of published papers recorded in the Web of Science. Across all science and engineering fields, the number of papers written by solo authors has remained relatively constant in absolute numbers but declined in terms of relative share of all papers written. By contrast, the size of authoring groups has increased each year. Pairs and trios were most frequent in the 1990–2000 period, while teams of 6 to 10 authors have been most common since 2000. Publications by very large groups of 100–1,000 authors first appeared in the 1980s, and publications by even larger groups of 1,000 or more authors appeared in the 2000s. The committee updated and analyzed the database, finding that, in 2013, about 95 percent of all papers were authored by 10 or fewer individuals, 5 percent were authored by 11 to 100 individuals, and less than 1 percent were authored by groups of more than 100 individuals.

Large numbers of participants can bring many benefits, yet also generate challenges, especially when the members are geographically dispersed.

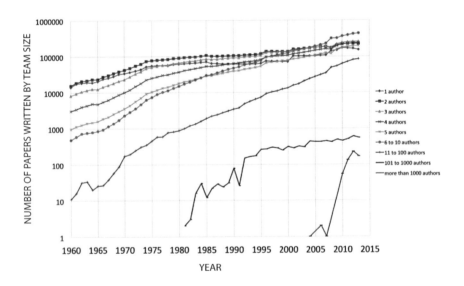

FIGURE 1-3 Frequency of author team sizes in science and engineering, 1960–2013.

For example, Stokols et al. (2008b) found that large multi-institutional team science projects are highly labor intensive, prone to conflict, and require substantial preparation and trust among team members to even partially achieve their scientific and translational objectives.

One large project that overcame such challenges was the U.S. effort to build an atomic bomb, known as the Manhattan Project. Initially, in 1941, small groups of physicists and engineers worked at their home universities. After Enrico Fermi demonstrated the first controlled nuclear reaction in 1942, the government built nuclear facilities at Oak Ridge, Tennessee; Hanford, Washington; and Los Alamos, New Mexico, ultimately employing 130,000 people. The scientific and military leaders overseeing the project faced the challenge of coordinating the tasks of thousands of production workers and motivating them to rapidly produce nuclear materials, while maintaining the secrecy of the project goal. In July 1945, scientists successfully detonated the world's first atomic bomb.

Although size is an increasingly important variable in the study of teams and groups, traditional research has rarely considered size to be a prime focus of analysis (Hackman and Vidmar, 1970; Stewart, 2006). Steiner (1972) identified the importance of team size as a determinant of a team's division of labor. By increasing team size, a problem is divisible into smaller parts along the line of "more hands make light work." Also, increasing team size could permit teams to effectively address larger-scale problems or more complex problems. For example, team size has been found to be positively related to the team-level recall of unique information, a driver of final performance of the team (Gallupe et al., 1992).

At the same time, larger team and group sizes are associated with process losses that can offset these potential benefits. As the number of members increases, the division of labor can become more inequitable (Liden et al., 2004) because of relational complexity and opportunities for "social loafing" if some members do little work (Latané, Williams, and Harkins, 1979). More generally, increases in group size require diverting time and resources from more productive activities to troubleshooting task interdependence, overcoming the tendency of individuals to "hoard" their unique knowledge, maintaining cooperative relationships, addressing incentive problems, and avoiding turnover (Jackson et al., 1991; Chompalov, Genuth, and Shrum, 2002; Okhuysena and Bechky, 2009).

Questions about the optimal size of groups remain open in part because the study of groups over time is difficult and in part because group size can have opposing effects on outcomes (e.g., a positive effect on productivity and a negative effect on cooperation). Recent work suggests that the effect of group size on productivity is moderated by the heterogeneity of the members. Observing the productivity of more than 549 information technology research teams and groups funded by NSF, it was found that larger groups were more productive 5 to 9 years later. Nevertheless, the marginal productivity declined as member heterogeneity rose, measured by increases in the number of disciplines included or the number of institutional affiliations (Cummings et al., 2013). This result reflects decades of research in the social sciences illuminating the challenge of dealing with the "other" and suggests that traditional forces such as ethnocentrism (whether applied to ethnic backgrounds or disciplinary perspectives) will continue to be a major challenge (Levine and Campbell, 1972).

As well as varying based on the degree of heterogeneity, it is likely that the challenges of large group size vary with the disciplinary context or cultural norms in which the team or larger group is embedded. For instance, the physics and genomics communities increasingly work in very large groups and publish with hundreds or even thousands of co-authors (Knorr-Cetina, 1999; Incandela, 2013). These scientific fields have developed infrastructures to support collaboration, including shared scientific instruments, data-sharing platforms, and publication guidelines and tools for large groups of collaborators (see Box 6-1 later in this report).

Goal Misalignment with Other Teams

Large groups of scientists, such as research centers and institutes, typically include multiple science teams engaged in research projects that are relevant to the higher-level goals of the center or institute. Each individual team brings valuable insights, methods, and perspectives and may have its own distinct goals. If the goals of these teams are not aligned, then this can generate conflict, requiring careful management.

Winter and Berente (2012) observed that research centers and other large groups are often composed of science teams from different organizations (e.g., disciplinary departments or medical centers) that may have conflicting or only partially overlapping goals. To some extent, goal misalignment with other teams is a natural consequence of including teams with diverse expertise and research or translational agendas. This problem is particularly salient in translational projects that involve teams of community stakeholders, such as policy makers and citizens, along with science teams. In such projects, it can be difficult for the component teams to formulate and reach consensus on shared, overarching goals, and the goals may change over time as the project evolves and the participants change (Cash, 2003; Hall et al., 2012a; Huutoniemi and Tapio, 2014).

The new concept of a "multiteam system," a complex system of interconnected teams, is beginning to increase understanding of the challenge of goal misalignment with other teams (Asencio et al., 2012). Such systems face the danger of "countervailing forces" that may advance goals at one level of the system, but slow collaboration at another level. One such force—strong cohesion within an individual team—may benefit that team's performance but may also discourage that team from sharing information with other teams that would benefit the system as a whole (DeChurch and Zaccaro, 2013). Furthermore, team members must balance devotion to the goals of their immediate team with the more distant goals of the broader organization or entity; the strong identification of members with a team can increase the success of the team, yet too strong an association with a proximal team can be at the expense of the higher order goals (DeChurch and Marks, 2006). For example, the James Webb Space Telescope, originally authorized in 1999, was expected in 2012 to cost nine times as much and to take a decade longer to complete than originally planned. The delays and cost overruns were attributed to inadequate budgeting for the inherent challenges of new technology development and weaknesses in managing and aligning the multiple academic and industry research and development teams engaged in the project (James Webb Space Telescope Independent Comprehensive Review Panel, 2010; U.S. Government Accountability Office, 2012).

Permeable Team and Group Boundaries

The boundaries of science teams and larger groups are often permeable, reflecting changes in the project goals and needs over time. The membership of a group or team may change as the project moves from one phase, requiring a certain type of expertise, to another that may require different expertise. Although these changes have the benefit of matching expertise to scientific or translational problems as they arise, they can also create challenges for effective team or group interaction.

Changes over time in the membership of a team or group may reflect the career stage and training needs of members as well as the research goals of the team or group. For example, studies of life sciences (Hackett, 2005) and physics laboratories (Traweek, 1988) have found that personnel turnover is ongoing, as students, postdoctoral fellows, and junior scientists are employed for a few years before moving on to other positions. However, unlike business employees who are typically assigned to work teams, scientists often voluntarily join science teams or groups. Therefore, scientists tend to have autonomy and operate like "free agents." A single scientist may belong to multiple teams at any one time, devoting more or less time to each one, depending on the level of funding available, the scientist's teaching and other research commitments, the potential for rewards, and other factors, including the scientist's personal interest in each particular project.

Permeable boundaries are central to transdisciplinary projects that blur not only disciplinary differences, but also the distinctions between scientists and lay people. It can be challenging to elicit lay knowledge in a form that can be integrated with formal expertise and applied to problems. Such challenges were overcome by the International Maize and Wheat Improvement Center in Mexico (Cash et al., 2003). Before the 1990s, center scientists had conducted research in laboratories or greenhouses to assure scientific rigor before transferring the resulting new crop breeds to the farmers. However, because the new crop breeds had sometimes lacked qualities desired by farmers and did not fit with existing crop management regimes, they were not widely adopted. By bringing the farmers directly into the project and working with them to find the most effective ways to integrate their unique knowledge, the center fostered more productive, sustainable farming practices on a wide scale (Cash, 2003).

The composition and number of team science projects with which a scientist is working can be in constant flux, creating potential challenges, as he or she juggles the conflicting time demands. One factor affecting a scientist's degree of involvement and allegiance to a particular team may be the level of knowledge integration involved. For instance, if a multidisciplinary project engages an expert briefly in a consulting role, the expert may not feel invested in, or identify with, the team. In contrast, an interdisciplinary or transdisciplinary research project may require all participants to engage more fully over a sustained period in order to integrate knowledge at a deeper level, fostering feelings of identity and investment in the shared work. Cummings and Haas (2012) found that teams whose members devoted a higher percentage of their time to the focal team performed more successfully than did teams whose members devoted a smaller percentage of their time to the focal team.

Teams in other contexts, including emergency response, the military teams, and business, also have permeable boundaries, with attendant benefits and challenges. For example, business teams engaged in new product development have permeable boundaries and changing membership, making it difficult to build trust and cohesion (Edmondson and Nembhard, 2009).

Geographic Dispersion

Most science teams and groups today are geographically dispersed. The dramatic increase in team-based research for more than four decades is due to collaborations that cross university boundaries. Jones, Wuchty, and Uzzi (2008) compared publications produced by solo authors, within-university groups, and multi-university groups each year from 1960 to 2005 across all U.S. institutions of higher learning. They found that while the publications by faculty from the same university remained flat since the 1950s, the increase in co-authored publications was due to the growth of groups from more than one institution.

Currently most scientists work virtually, even with co-located colleagues, but the further geographically dispersed participants are across locations and institutions, the greater the possibility for coordination and communication challenges to emerge. Working across long distances introduces challenges such as a limited number of overlapping work hours among members located in different time zones and differences in incentives structures for members in different organizations. As noted above, some scientists' identity is closely related to the particular work styles, technologies, and routines of their particular laboratories (Knorr-Cetina, 1999; Hackett, 2005).

Science teams or groups including scientists from different institutions and perhaps different countries may find it difficult to foster shared identification with the project, and to develop common work styles. Additionally, questions regarding access to technology and data can generate challenges. For example, the Thirty Meter Telescope is currently being developed by a large scientific organization including the California Institute of Technology, University of California System, National Astronomical Observatories of the Chinese Academy of Sciences, and National Institute of Natural Sciences/National Astronomical Observatory of Japan. The involvement of scientists from nations with different languages, cultures, politics, and economies could potentially lead to misunderstandings or conflicts.

Teams in business, the military, and other sectors are also increasingly geographically dispersed (Kirkman, Gibson, and Kim, 2012), gaining the benefits of global expertise and encountering similar challenges. Chapter 7 discusses the benefits and challenges of geographically distributed work and provides strategies and recommendations for remediating the challenges.

High Task Interdependence

One of the defining features of a team is that the members are dependent on each other to accomplish a shared task, and science teams are no exception (Kozlowski and Ilgen, 2006; Fiore, 2008). All team science projects, regardless of size or level of disciplinary integration, face challenges related to effectively developing and conducting a shared research agenda. The process of designing and conducting interdependent tasks that draw on and integrate the unique talents of the individual members is challenging, but such interdependence is the norm among the very large groups of physicists who conduct research at the Large Hadron Collider in Geneva, Switzerland. Knorr-Cetina (1999) suggested that the interdependence is inherent in the nature of research that can only be conducted at a few very large sites, leading to a unique "communitarian" culture in high-energy physics (see Box 6-1 for further discussion).

Fiore (2008) proposed that scientists engaged in interdisciplinary and transdisciplinary research projects are more interdependent than those involved in team science projects that do not require a high degree of knowledge integration. He noted that some scientists avoid interdisciplinary research because they believe they must master multiple disciplines, yet teams in organizations are brought together to achieve shared goals with the recognition that the team members will only be able to develop partially overlapping knowledge.

Greater task interdependence among team or group members can lead to more opportunities for conflicts. Furthermore, when geographically dispersed colleagues must perform highly interdependent tasks, greater coordination and communication efforts may be required to bridge boundaries and facilitate task completion. The challenges of task interdependence and research-based strategies to address these challenges are discussed in Chapters 3 and 4. The unique challenges of task interdependence in dispersed teams are addressed in Chapter 7.

LEARNING FROM RESEARCH ON TEAMS IN OTHER CONTEXTS

Research on teams in contexts outside of science provides a rich foundation of knowledge about team processes and effectiveness. Because teams in science share features and processes with teams in other contexts, and based on the history of generalization of team research across contexts, the committee assumes that this knowledge can inform strategies for improving the effectiveness of science teams and larger groups. Below, we elaborate on these points.

Similar Features

Much of the existing scientific literature about teams has focused on teams in contexts outside of science, such as the military, factories, intelligence analysis, medicine, and emergency response. These teams in other contexts increasingly share the seven features that can create challenges for team science.

In medicine, for example, patient care is carried out by teams of physicians, nurses, and technicians with diverse specialties, who experience the benefits and challenges of high diversity in team membership as they seek to combine their knowledge to effectively solve problems. Intelligence analysts filter and fuse information to make decisions, solve problems, or create new knowledge, as do project teams and research and development teams (Heuer, 1999; Kerr and Tindale, 2004). All of these teams in other contexts seek to deeply integrate their knowledge, as do interdisciplinary and transdisciplinary science teams. In terms of size, teams in these other contexts are similar to science teams, typically including 10 or fewer members.

In the military, corporations, and health care, leaders are replacing traditional departments and divisions with agile teams that have permeable boundaries, adding members when particular skills or expertise are needed, while losing members who are assigned to different teams (Tannenbaum et al., 2012). Corporations once divided into departments specializing in research and development, sales, and production are creating temporary new product development teams that combine all of these functions. Edmondson and Nembhard (2009) identified five features of new product development teams that simultaneously offer the potential for innovation and pose challenges; several of these features also create challenges for team science. They include (1) project complexity, (2) cross-functionality, (3) temporary membership, (4) fluid team boundaries, and (5) embeddedness in organizational structures. The authors emphasized that successfully managing these features yields both organization-level benefits and team-level benefits, in the form of new capabilities and team member resilience.

Businesses with multiple, agile teams face the challenge of goal misalignment with other teams, and their employees and executives face the challenge of juggling the demands of the multiple teams (Espinosa et al., 2003; O'Leary, Mortenson, and Woolley, 2011; Maynard et al., 2012). Teams in business, governmental organizations, and in many other contexts are increasingly geographically dispersed, relying more heavily than in the past on technology to support their communication (Kirkman, Gibson, and Kim, 2012).

All of these features (highly diverse membership, deep knowledge integration, large size, goal misalignment, permeable boundaries, geographic dispersion, and high task interdependence) create challenges for science teams and teams in other contexts.

Similar Processes

Research in other contexts has demonstrated that certain interpersonal processes within teams, such as conflict, cohesion, and shared understanding of goals, are related to achieving team goals (Kozlowski and Ilgen, 2006; see Chapter 3). This research has also illuminated approaches that can be used by team leaders and members to influence these processes in positive ways, thereby increasing team effectiveness (i.e., performance). Recent research focusing specifically on science teams and groups has begun to yield similar findings about the importance of interpersonal processes. For example, intellectual conflicts and disagreements are important processes for advancing knowledge in science and other fields (Collins, 1998). Bennett and Gadlin (2012) analyzed in-depth interviews with members of successful science teams and others that ended because of conflict or did not meet their goals. They found that the more successful teams promoted intellectual disagreement and discussion—which brought such benefits as continuing the dialogue, working through issues, and keeping problems or issues from accumulating—while also containing conflict and developing trust. In another example, the research on teams in non-science contexts has demonstrated that leadership styles and behaviors can positively influence interpersonal team processes, thereby improving team performance (Kozlowski and Ilgen, 2006). Similarly, a study of research laboratories in Europe found that the quality of laboratory directors' supervision was positively related to the working climate and research productivity of the laboratories they directed (Knorr et al., 1979).

Generalizing the Research Across Contexts

Teams have been studied in a variety of organizational contexts, and findings in one context have often been generalized to other contexts. For instance, guided team self-correction, also known as team dimensional training, is a research-based approach that helps a team reflect on its teamwork during a past performance episode, identify errors, and develop solutions (Smith-Jentsch et al., 2008; see Chapter 5). It has been shown to improve performance in Navy attack center and shipboard teams and has been generalized to augment teamwork simulation exercises for Navy aircrews, engineering, seamanship, damage control, and combat systems teams, as well as civilian firefighting teams, law enforcement teams, and teams of corrections officers. Finally, it has been used as a tool to support on-the-job performance improvement through accident investigations within the nuclear power industry and to debrief one organization's response to the terrorist attacks of 9/11 (Smith-Jentsch et al., 2008). Because guided team self-correction is based on a model of expert teamwork behaviors within a particular organizational context, the approach was translated to each new context based on analysis of the components of expert teamwork in that context.

Another example, Crew Resource Management training, was developed in the aviation industry to improve air travel safety by increasing teamwork and communication and reducing human error in the cockpit. The approach is widely used in the airline industry, has gained acceptance from airline crews, and has been shown to change crew behaviors (Helmreich, Merritt, and Wilhelm, 1999; Pizzi, Goldfarb, and Nash, 2001). Crew Resource Management provided the basis for guided team self-correction training described above and has also been translated for health care in TeamSTEPPS training. TeamSTEPPS is designed to improve patient safety by increasing communication and decreasing medical errors within patient care teams (King et al., 2008).

Therefore, based on the similarities in features and processes between teams in science and those in other contexts and the history of generalization of team research across contexts, the committee assumes that research on teams in other contexts provides a rich foundation of knowledge about team processes and effectiveness that can inform strategies for improving the effectiveness of science teams and larger groups.

THE VALUE OF MULTIPLE APPROACHES AND THE PROMISE OF TEAM SCIENCE

Although team science is growing rapidly, individual scientists continue to make critical contributions and important discoveries, as exemplified by Stephen Hawking's stream of new insights into the nature of the universe. Public and private funders with finite budgets must make decisions about whether to develop individual investigator or team approaches, and, if a team approach is selected, the scale and scope of the project. Similarly, individual scientists must make decisions about whether to invest time and energy in collaborative projects or to focus on individual investigations. It is important for scientists and other stakeholders to strategically consider the particular research question, subject matter, and intended scientific and/ or policy goals when determining whether a team science approach is appropriate, and if so, the suitable size, duration, and structure of the project or projects (Westfall, 2003).

Several strands of research and data suggest that team science can rapidly advance scientific and technological innovation by increasing research impact, novelty, productivity, and reach. First, group publications are more highly cited than publications by individuals, an indicator of their impact. Wuchty, Jones, and Uzzi (2007) found that teams and groups typically produce more highly cited publications and patents than do individuals (even eliminating self-citations), and that this advantage is increasing over time. Second, Uzzi and colleagues (2013) found evidence of both impact and novelty in team science: Compared with solo authors, teams and groups across disciplines were more likely to put novel combinations of prior work together, and to develop work that assimilated novel ideas into high-impact publications. Third, in a quasi-experimental comparative study, Hall et al. (2012b) found that transdisciplinary tobacco use research centers (large science groups) had higher overall publication rates and published findings from funded projects more consistently than did individuals or small teams investigating tobacco use, highlighting benefits for research productivity and dissemination. Fourth, Stipelman and colleagues (2014) compared the structure and disciplinary topical coverage of publications over time of transdisciplinary research centers with those of two comparison groups consisting of individuals and small teams. An overlay of the resulting publication data on a base map of science revealed that the publications from the transdisciplinary research centers spread across the disciplinary topics in the map of science more rapidly and more comprehensively than both comparison groups, suggesting that the transdisciplinary team science approach broadens the reach of research findings across areas of science. Finally, the rapid growth of co-authored publications since 1960 documented by Wuchty, Jones, and Uzzi (2007) reflects the expert judgment of scientists in research funding agencies and peer review panels that teams or larger groups were best suited to address important research questions and that the results were worthy of publication.

In light of both the rapid growth and promise of team science, and the seven features that can create challenges, funding agencies and policy makers need to identify the most effective strategies for ensuring that taxpayer investments in team science yield valuable returns (Croyle, 2008, 2012). Scientists and leaders of teams and groups also need information on how to effectively manage these projects. The first step toward increased effectiveness is to gain understanding of the factors that facilitate or hinder team science and how these factors can be leveraged to improve the management, administration, and funding of team science. Although research is emerging from the science of team science, the research on teams, and from many other fields, it is fragmented, and team science practitioners may find it difficult to access or to understand and apply. This report integrates and translates the relevant research to support conclusions and recommendation for practice and identify areas requiring further research.

STUDY APPROACH

The NRC convened a Planning Meeting on Interdisciplinary Science Teams in January 2013 to raise awareness of this study, begin to explore the relevant literature, and solicit input from federal agencies, individual investigators, team science researchers, directors of research institutions, and other stakeholders (see http://tvworldwide.com/events/nas/130111/# [April 2015]).

The NRC then convened this committee, which met for the first time in April 2013. At its April meeting, the committee heard presentations from current and former NSF officials about the need for the study and from psychologist Gregory Feist, who focused on scientific creativity. Most of the meeting was spent in closed session discussing the study charge and how to approach it. The committee's second meeting, in July 2013, included a Workshop on Team Dynamics and Effectiveness, which explored many individual-level and team-level factors that influence the processes and outcomes of team science (see http://www.tvworldwide.com/events/

nas/130701/ [April 2015]). The committee's third meeting, in October 2013, included a Workshop on Organizational and Institutional Supports for Team Science. Speakers at this workshop included researchers who study organizational factors and university leaders with practical knowledge of how to support team science. The committee's fourth meeting, a virtual meeting, focused primarily on draft chapters, conclusions, and recommendations of the consensus report and also included a brief discussion with the NSF study sponsors. The committee met for its fifth and final time in March 2014. At this meeting, the committee reached consensus on its conclusions and recommendations and discussed finalizing this report.

ORGANIZATION OF THE REPORT

This report is designed to address the guiding questions in the committee charge (Box 1-1). It is organized into four parts, as follows:

- **Part I: Setting the Stage.** Chapters 1 and 2 provide the key definitions and conceptual framework for the research review in Parts II and III.
- **Part II: The Individual and Team Levels.** Chapter 3 provides an overview of the research on team effectiveness. It identifies team process factors at the individual and team levels and ways to manipulate three aspects of a science team or larger group to enhance effectiveness—its composition, professional development, and leadership. The following three chapters address each aspect, focusing in turn on team composition (Chapter 4), professional development and education (Chapter 5), and team and organizational leadership (Chapter 6).
- **Part III: The Institutional and Organizational Level.** Chapter 7 discusses the challenges of geographically distributed science teams and larger groups, and the role of organizations, leaders, and cyber infrastructure in addressing these challenges. Chapter 8 discusses organizational support for team science, focusing particularly on research universities. Chapter 9 considers the role of funding organizations that provide financial and other supports for team science.
- **Part IV: A Path Forward.** Chapter 10 provides a research agenda to advance research on team science effectiveness.

Reflecting the complex, multifaceted nature of team science and the multiple levels of analysis required to begin to understand it, many questions in the study charge are addressed in more than one chapter. For example, the role of individual characteristics in science team effectiveness is introduced in Chapter 3 and discussed in greater detail in Chapter 4. Similarly, leadership influences team science not only at the level of the team, but also at the level of the research organization and the funding agency, often expressed in the development of "structures, policies, practices, and resources." Hence, issues related to management and leadership are introduced in Chapter 3, elaborated upon in Chapter 6, and also discussed in Chapter 8. Table 1-3 depicts the coverage of the questions in the committee's charge in the report chapters.

TABLE 1-3 Coverage of the Charge in the Report

Chapter	Questions in the Study Charge
Chapter 1: Introduction	
Chapter 2: Science to Inform Team Science	
Chapter 3: Overview of the Research on Team Effectiveness Chapter 4: Team Composition and Assembly Chapter 5: Professional Development and Education for Team Science Chapter 6: Team Science Leadership	1. How do individual factors (e.g., openness to divergent ideas) influence team dynamics (e.g., cohesion), and how, in turn, do both individual factors and team dynamics influence the effectiveness and productivity of science teams?
Chapters 1, 3, and 4 Chapter 7: Supporting Virtual Collaboration	2. What factors at the team, center, or institute level (e.g., team size, team membership, geographic dispersion) influence the effectiveness of science teams? 5. What factors influence the productivity and effectiveness of research organizations that conduct and support team and collaborative science, such as research centers and institutes? How do such organizational factors as human resource policies and practices and cyber infrastructure affect team and collaborative science?
Chapters 4 and 6	1. How do individual factors (e.g., openness to divergent ideas) influence team dynamics (e.g., cohesion), and how, in turn, do both individual factors and team dynamics influence the effectiveness and productivity of science teams? 3. How do different management approaches and leadership styles influence the effectiveness of science teams?

TABLE 1-3 Continued

Chapter	Questions in the Study Charge
Chapter 8: Institutional and Organizational Support for Team Science	4. How do current tenure and promotion policies acknowledge and provide incentives to academic researchers who engage in team science?
	5. What factors influence the productivity and effectiveness of research organizations that conduct and support team and collaborative science, such as research centers and institutes? How do such organizational factors as human resource policies and practices and cyber infrastructure affect team and collaborative science?
	6. What types of organizational structures, policies, practices, and resources are needed to promote effective team science, in academic institutions, research centers, industry, and other settings?
Chapter 9: Funding and Evaluation of Team Science	5. What factors influence the productivity and effectiveness of research organizations that conduct and support team and collaborative science, such as research centers and institutes? How do organizational factors such as human resource policies and practices and cyber infrastructure affect team and collaborative science?
	6. What types of organizational structures, policies, practices, and resources are needed to promote effective team science, in academic institutions, research centers, industry, and other settings?
Chapter 10: Advancing Research on the Effectiveness of Team Science	All questions

2

Science to Inform Team Science

The preceding chapter defined "team science" as scientific collaboration conducted by more than one individual in an interdependent fashion. It also identified seven features that create challenges for team science. This chapter focuses on two of the scientific fields that have centrally contributed diverse methodological and conceptual approaches to understanding and addressing these challenges. Together, these fields provide cumulative empirical knowledge to assist scientists, administrators, funding agencies, and policy makers in improving the effectiveness of team science. We first discuss the social science research on groups and teams and then the "science of team science," an emerging, interdisciplinary field focusing, as its name suggests, specifically on team science.

RESEARCH ON GROUPS AND TEAMS

This report draws heavily from the social science literature of groups and teams. Organizational, cognitive, and social psychologists have studied team processes and outcomes for more than four decades, providing strong evidence about processes that enhance team performance and how those processes can be influenced (e.g., Kozlowski and Ilgen, 2006; Mathieu et al., 2008; Salas, Cooke, and Gorman, 2010; see also Chapter 3 in this report). As noted in the previous chapter, much of this research focuses on teams in contexts outside of science, yet these teams in other contexts incorporate many of the key features that create challenges for team science. In addition, emerging research focusing specifically on science contexts is beginning to identify similar processes to those identified in other contexts. Thus,

47

this research is relevant to science teams, and we draw extensively on it in Chapters 3 through 6. In addition, some studies have focused specifically on industrial research and development teams, which are typically composed of scientists engaged in research, similar to academic science teams. For example, Bain, Mann, and Pirola-Merlo (2001) examined the relationship between team climate and performance in research and development teams, and Keller (2006) studied leadership in research and development product teams.

Research on groups and teams has benefitted from the use of simulation and modeling, and it is likely that research on team science can benefit similarly. Simulation allows technological tasks conducted by science teams in the real world (e.g., joint use of scientific equipment or virtual meeting technologies) to be studied under controlled laboratory conditions (Schiflett et al., 2004). For instance, simulation can be used to mock up technologies that human users interact with in the laboratory. One or more technologies can then be evaluated on usability as well as on their ability to improve effectiveness in a science team or group. In addition, agent-based modeling, dynamical systems modeling, social network modeling, and other forms of computational modeling have become more prevalent in the teams literature and can help to extend empirical results from small science teams to larger groups of scientists and scientific organizations (National Research Council, 2008; Gorman, Amazeen, and Cooke, 2010; Kozlowski et al., 2013; Rajivan, Janssen, and Cooke, 2013).

THE SCIENCE OF TEAM SCIENCE

The complex and variegated nature of team science makes the scientific investigation of all its dimensions and contexts quite challenging. Toward the goal of better understanding these inherent complexities, a new field, the science of team science, has emerged (e.g., Croyle, 2008; Stokols et al., 2008a; Fiore, 2008, 2013). In this chapter, we identify some of the unique concerns and contours of this rapidly expanding field, which has been defined as:

> a new interdisciplinary field . . . which aims to better understand the circumstances that facilitate or hinder effective team-based research and practice and to identify the unique outcomes of these approaches in the areas of productivity, innovation, and translation. (Stokols et al., 2013, p. 4)

While drawing heavily on the perspectives and findings from research on groups and teams, scholars in the science of team science are concerned with a number of questions that have not been addressed explicitly in that research, as discussed below.

Distinctive Concerns of the Science of Team Science

The scholarly and applied concerns of the science of team science are closely related to the seven features outlined in Chapter 1 that can pose challenges. The distinctive concerns of the field include

- focusing on highly diverse units of analysis, ranging from the level of the team to broader organizational, institutional, and science policy contexts, including centers and institutes specifically designed to promote and sustain team science;
- understanding the multinetwork structure of scientific collaboration, including the diverse contexts and pathways of collaboration that have emerged in recent years;
- understanding the promise and challenges of diverse team membership and deep knowledge integration, especially in transdisciplinary projects that aim to achieve practical as well as scientific innovations;
- establishing reliable, valid consensus criteria for evaluating team science processes and outcomes; and
- focusing on translational and educational as well as scientific goals.

Focusing on Highly Diverse Units of Analysis

Team science encompasses an enormously diverse set of arrangements for conducting collaborative science. As discussed in the previous chapter, team science projects vary in size, duration, level of funding, geographic dispersion, and level of disciplinary integration (Stokols, 2013). Reflecting this diversity, the field focuses on multiple, interacting levels, posing challenges for theory and research.

First, at a team level of analysis, the science of team science field focuses on science teams and groups and their individual members as the principal units of study. Chapter 3 reviews various individual- and team-level factors that influence the functioning and outputs of science teams and larger groups.

As the field's focus moves beyond individual science teams to higher levels of analysis, it focuses on a variety of organizations and institutions whose mission or goals are to facilitate and sustain effective team science collaboration (Börner et al., 2010; Falk-Krzesinski et al., 2011). For example, universities often establish new research centers focusing on particular scientific and societal problems (e.g., cancer control and prevention; environmental sustainability) to facilitate cross-disciplinary team-based research addressing these problems. Such centers often support several different science teams that may work together in pursuit of shared research goals as part of a multiteam system (DeChurch and Zaccaro, 2013).

In addition to its special focus on organizations such as research centers, the science of team science seeks to understand more generally the extent to which various scientific organizations and institutions (e.g., research universities, national laboratories, research funding agencies) may support or hinder team science (see Chapter 8 for further discussion). For example, researchers might analyze how research university incentive structures, such as promotion and tenure policies, affect scientists' motivation to participate in team science. As another example, one recent study assessed the relative scientific productivity rates of tobacco scientists participating in National Cancer Institute Transdisciplinary Tobacco Use Research Centers (TTURCs) with those of National Institutes of Health grantees working on the same topics as members of smaller research teams who are not participating in the broader research centers (Hall et al., 2012b). Such questions about the effectiveness of alternative research infrastructures or the translational impacts of team science programs have not been explicitly addressed in earlier research on non-science teams.

Finally, at the broadest level of analysis, the field is concerned with how community and societal factors, including social, cultural, political, and economic trends, influence decisions to use a team science approach, the selection of phenomena to be investigated, and the prospects for successful collaboration in the investigation (Institute of Medicine, 2013). For example, policy makers, health care professionals, and scientists are currently focused on ameliorating the national trend of increasing obesity with its attendant adverse health effects (e.g., Institute of Medicine, 2010). Here, science policy concerns rise to the fore, as researchers study the design of funding mechanisms to encourage and sustain science teams and groups, as well as peer review and program evaluation criteria (e.g., Holbrook, 2013; Jordan, 2013) for judging the effectiveness of such teams and groups (see Chapter 9 for further discussion).

Understanding the Multinetwork Structure of Contemporary Scientific Collaboration

Social scientists have begun to investigate the important role of networks in advancing scientific knowledge. For example, sociologist Randall Collins (1998) conducted a comprehensive sociological analysis of the intellectual debates and relationships within and among networks of scholars since the time of the ancient Greeks, arguing that these networks have catalyzed major intellectual advances in philosophy, science, and other fields. In another example, Mullins (1972) traced the creation of molecular biology as a new scientific discipline in the 1960s to the evolving networks of relationships among a group of colleagues, students, and co-authors studying the bacteriophage, a virus that infects bacteria.

Today, team science increasingly takes place through multiple networks and teams that may be closely linked or unrelated. A given scientist may participate to varying degrees in these networks and teams. The science of team science field is concerned with understanding this multinetwork structure of scientific collaboration in the early 21st century (Shrum, Genuth, and Chompalov, 2007; Dickinson and Bonney, 2012; Nielsen, 2012). Scientists often simultaneously participate in multiple teams, and these teams are embedded within larger networks that are based on their past collaborations (Guimera et al., 2005). These large scientific and translational networks include closely linked groups of individuals who have conducted research and perhaps published together, and also more loosely affiliated groups. For example, some members of the committee that authored this report previously collaborated extensively with others to evaluate National Cancer Institute team science projects (e.g., Stokols, Hall, and Vogel, 2013), others are affiliated through their shared membership in the National Academy of Sciences, and still others are affiliated as faculty members at the same universities.

Understanding the Promise and Challenges of Diverse Membership and Deep Knowledge Integration

Another complexity facing the science of team science is to understand and address the communication and coordination challenges emerging from the first two features that pose challenges for team science introduced in Chapter 1—high diversity of team or group membership and deep knowledge integration. The challenges are especially great in transdisciplinary projects that may have multiple scientific and societal goals and require high levels of knowledge integration across disciplines and professions (Frodeman et al., 2010). Thus, a critical issue for the science of team science involves examination of the integrative processes and outcomes in disciplinarily heterogeneous science teams and how they lead to scientific innovations. This understanding is needed whether the project aims for "translational" innovations that are more immediately applicable or more fundamental scientific knowledge.

Establishing Reliable, Valid, and Consensual Criteria for Evaluating Team Science Processes and Outcomes

Evaluating the processes and outcomes of science teams and groups is particularly challenging because of their multiple goals. As the research focus of the science of team science shifts from small, short-term science teams to larger, more enduring organizational and institutional structures, the goals of a project and the criteria for judging its success vary accordingly. Whereas the primary goals of small teams may entail the creation and dissemination of new scientific knowledge, larger team science centers and institutions often encompass broader goals. Reflecting their multiple goals, large organizational structures require broad metrics to evaluate their effectiveness. Such metrics may include assessments of the extent to which the smaller team science projects they administer bring about intellectual innovations in the near term, and the extent to which the organization is able to coordinate and integrate across projects to translate these near-term scientific findings into new technologies, policies, and/or community interventions (i.e., scientific and societal returns—see Chapter 9 for further discussion).

These higher-level organizations and institutions (e.g., a research center or institute) must be responsive to the scientific and translational priorities embraced by their community and governmental funders, whereas these priorities may be much less salient to individual scientists working on individual projects (Winter and Berente, 2012). Thus, an important concern of the science of team science field is to develop evaluative criteria that are appropriately matched to the respective goals and concerns of the teams, organizations, institutions, funders, and community groups that have a stake in the foci, processes, and outcomes of large programs of team science research. Scholars in the science of team science are concerned with the relative efficacy of alternative team science funding mechanisms and the development of criteria for evaluating the returns on investments in team science projects—questions that have not been explicitly addressed in earlier research on non-science teams (Winter and Berente, 2012).

The field is also increasingly concerned with articulating appropriate criteria for measuring the potential (ex-ante) and achieved (ex-post) outcomes of science teams and larger groups, including those that focus within a single discipline and those that cross disciplines (Holbrook, 2013; Jordan, 2013; Stokols, 2013). In particular, a growing number of science teams and groups have transdisciplinary goals, seeking to achieve scientific advances by not only integrating, but also transcending multiple disciplinary perspectives and to apply the resulting scientific advances (Croyle, 2008; Crow, 2010; Klein, 2010). In response to this trend, the field is concerned with identifying reliable, valid, and consensually agreed-upon criteria for judging the success of such transdisciplinary projects relative to those that are uni- or multidisciplinary (Frodeman et al., 2010; Pohl, 2011).

As a first step toward developing such criteria, the field must develop measures of the processes leading to effectiveness. As teams and groups develop and move through their phases of scientific problem solving, their interactions will change, and the field must identify how to measure these team processes. Such measures will aid understanding of how team processes are related to the multiple goals of transdisciplinary team science projects. Achieving this understanding requires articulation of a comprehensive, multimethod measurement approach that includes, but is not limited to, bibliometric indices, co-authorship network analyses, experts' subjective appraisals of team science processes and products, and surveys and interviews of team science participants. Particularly challenging is the measurement of deep interdisciplinary knowledge integration (Wagner et al., 2011), but there are new methods and measures that appear promising as discussed in Chapter 9. Such efforts to measure team processes are often more daunting than developing evaluative criteria to measure team outcomes in other settings. As part of this measurement challenge, the field needs to more clearly differentiate the processes and outcomes of unidisciplinary, multidisciplinary, interdisciplinary, and transdisciplinary science teams.

An essential first step in the process of establishing evaluative criteria is to gain access to practicing scientists to study their interactions and innovations. Although some funding agencies and scientists themselves resist providing such access, it is critical for advancing the science of team science. For example, more than a decade ago, the Institute of Medicine (1999) produced a groundbreaking report on patient safety and errors in health care. As a result, researchers began to gain access to health care settings, illuminating the relationship between medical teams' processes and patient outcomes and identifying strategies for reducing errors and improving patient safety (e.g., Edmondson, Bohmer, and Pisano, 2001). Providing researchers access to science teams embedded in their research contexts promises similar benefits.

Focusing on Translational and Educational as Well as Scientific Goals

Finally, the science of team science field is concerned with not only research but also translation of the research to improve practice (Spaapen and Dijstebloem, 2005; Stokols et al., 2008a). The translational goals of the field include

- using the research findings on team science to improve community and societal conditions (e.g., through the development of improved clinical practices, disease-prevention strategies, public health policies);
- applying research findings from evaluations of large team science research projects to improve future scientific teamwork and designing organizational, institutional, educational, and science policies

and practices to promote effective team science (see further discussion in Chapters 3–9); and

- developing education and training programs and resources to enhance students and scholars' capacity for effective scientific collaboration in their future or current team science endeavors (Stokols, 2006; COALESCE, 2010; Klein, 2010; National Institutes of Health, 2010; National Cancer Institute, 2011; Vogel et al., 2012; see also Chapter 5).

A Complex Adaptive Systems Approach

Researchers have begun applying the methods and perspectives of complexity science to help understand and address the communication and coordination challenges of team science.

Complexity science uses computer simulations to study "complex adaptive systems," which are systems made up of multiple parts that continually interact and adapt their behavior in response to the behavior of the other parts (Holland, 1992). By modeling such systems, researchers seek to understand how the aggregate behavior of the system emerges from the interactions of the parts, integrating multiple levels of analysis to build a more thorough understanding of phenomena. For example, Liljenström and Svedin (2005) described a complex adaptive system as a network of non-linear interactions within an open system, which produce a form of self-organization and emergence. It may be relevant to draw on complexity theory to bound the levels of analysis and address the theoretical and measurement issues present in team science environments. Organizational scientists refer to team effectiveness as "emergent," because it originates in the thinking and behaviors of individual team members and is amplified by team members' interactions (Kozlowski and Klein, 2000). Kozlowski et al. (2013) have studied emergent collaboration, and Kozlowski et al. (in press) have examined knowledge emergence in decision-making teams, relevant to the challenge of deep knowledge integration in interdisciplinary and transdisciplinary science teams (Kozlowski and Klein, 2000; see further discussion in Chapter 3).

By virtue of their multiple levels of scale (individual, team, organizational, multi-institutional) and many different actors with various motivations and priorities, science teams and groups can display the major characteristics of a complex adaptive system as described by Hammond (2009). Börner et al. (2010) called for a multilevel systems perspective to advance the science of team science. This approach would include macro-level analyses to help understand broad patterns of collaboration within and across scientific fields (e.g., Klein, 1996), meso-level analyses to understand the social and group processes arising during collaboration in science teams and groups (e.g., Fiore, 2008), and micro-level analyses to understand the individuals that comprise the science teams (e.g., their education and training, their motivation). Similarly, Falk-Krzesinski et al. (2011) cautioned that "sequential process models could not adequately capture the complexity inherent in SciTS [the science of team science] and may even be misleading" (p. 154). They argued that a systems view is more appropriate as it can help better account for interdependence and the iterative relationships among the components of science teams and the contexts in which they operate.

OTHER CONTRIBUTING FIELDS OF RESEARCH

Many other fields of research in addition to the science of team science and the research on groups and teams contribute to an understanding of team science and how to increase its effectiveness. These include social studies of science (e.g., Galison, 1996), science and technology studies (Pelz and Andrews, 1976), history and philosophy of science, cultural anthropology, and organizational and management studies (e.g., Kellogg, Orlikowski, and Yates, 2006), as well as interdisciplinary studies, information science, the humanities, and program evaluation research. A detailed examination of the contribution of these fields is beyond the scope of this report, but we provide some examples of relevant work in this section.

Sociologists and economists have examined the internal and external forces motivating individual scientists. For example, sociologist Robert Merton (1968) found that well-known scientists were given disproportionate credit for collaboratively authored publications, increasing their visibility while reducing the visibility of less well known contributors. Social scientists continue to study how credit and rewards are allocated when scientists collaborate (e.g., Furman and Gaule, 2013; Gans and Murray, 2015) revealing tensions that affect scientists' willingness to join science teams and groups (see Chapter 8 for further discussion).

Anthropologists and sociologists have conducted in-depth studies of scientific laboratories in the life sciences, high-energy physics, and other disciplines (e.g., Latour and Woolgar, 1986; Knorr-Cetina, 1999; Owen-Smith, 2001; Hackett, 2005). Cognitive scientists have also conducted studies of scientific work in particular settings, while psychologists have examined the role of scientists' personality characteristics and other factors in supporting scientific creativity and productivity (e.g., Simonton, 2004; Feist, 2011, 2013). Building on studies focusing on individual scientists, recent research has begun to explore collaborations between scientific institutions (e.g., Shrum, Genuth, and Chompalov, 2007; Garrett-Jones, Turpin, and Diment, 2010; Bozeman, Fay and Slade, 2012; see Chapter 8 for further discussion).

SUMMARY

In this chapter, we have described several fields that contribute to understanding how to improve the effectiveness of team science. This report draws heavily on the robust literature from research on groups and teams and on the body of research emerging from the science of team science. We have described the interdisciplinary and multilevel orientation of the science of team science and outlined several of its distinctive challenges and concerns. Many other fields contribute to the committee's understanding of the effectiveness of team science, but are beyond the scope of this report, including social studies of science, organizational and management studies, industrial-organizational and cognitive psychology, science and technology studies, interdisciplinary studies, communications and information science, the humanities, and program evaluation research.

Part II

The Individual and Team Levels

3

Overview of the Research
on Team Effectiveness

This chapter summarizes the research literature on team effectiveness, highlighting findings on the key features that create challenges
for team science outlined in Chapter 1. Based on its review of the
literature (e.g., Marks, Mathieu, and Zaccaro, 2001; Kozlowski and Ilgen,
2006; Salas, Goodwin, and Burke, 2009), the committee defines team effectiveness as follows:

> Team effectiveness, also referred to as team performance, is a team's ca-
> pacity to achieve its goals and objectives. This capacity to achieve goals
> and objectives leads to improved outcomes for the team members (e.g.,
> team member satisfaction and willingness to remain together) as well as
> outcomes produced or influenced by the team. In a science team or larger
> group, the outcomes include new research findings or methods and may
> also include translational applications of the research.

More than half a century of research on team effectiveness (Kozlowski
and Ilgen, 2006) provides a foundation for identifying team process factors
that contribute to team effectiveness, as well as actions and interventions
that can be used to shape the quality of those processes. As noted in Chapter 1, this evidence base consists primarily of studies focusing on teams in
contexts outside of science, such as the military, business, and health care.
These teams share many of the seven features that can create challenges for
team science introduced in Chapter 1. For example, in corporations, top
management teams and project teams are often composed of members from
diverse corporate functions, and these teams seek to deeply integrate their
diverse expertise in order to achieve business goals. Therefore, the commit-

BOX 3-1
What Is a Meta-Analysis?

The foundation of scientific research is based on primary studies that collect data under a given set of conditions (i.e., experiments or field studies) and examine effects on, or relationships among, the observed variables of interest. However, all research is subject to limitations and no single study is definitive. Thus, there is considerable value in the use of a meta-analysis to quantitatively combine multiple primary studies and summarize their findings. The basic steps of a meta-analysis include (a) conducting a thorough search for relevant studies (including unpublished ones); (b) converting test statistics to effect sizes (i.e., an index capturing the strength of the relationship between two variables); (c) weighing the effect size from a study by its sample size (i.e., studies with larger samples presumably contain less-biased estimates of the true effect size and therefore receive higher weight); and (d) combining the effect sizes across studies to estimate the overall strength and meaningfulness of a given relationship (i.e., testing for statistical significance and establishing confidence intervals). Depending on the number, scope, and sample size of the primary studies, the average effect size can be generalized as a population estimate of the relationship in question. In addition, a meta-analysis often corrects the raw averaged effect size for a variety of statistical artifacts (i.e., measurement unreliability, restriction of range from sampling) to improve population effect size estimate.

Depending on what it is possible to code from the primary studies, a meta-analysis may examine other factors that moderate or change the strength of a relationship (e.g., whether the research was experimental or field based; whether it was one type of team vs. another type of team).

Effect sizes can be reported using a variety of indices, but r (i.e., correlation) is often used for uncorrected effects and ρ (i.e., rho) for corrected ones. The interpretation of r and ρ is straightforward. The indices range from -1.00 to +1.00 to indicate the strength and direction of the relationship. Cohen's (1992) Rules-of-Thumb designate correlations (r) of .10 as small, .25 as medium, and .40 as large effect sizes. Squaring the two indicators gives a direct measure of the proportion of variance shared by both variables. Thus, an effect size of .35 accounts for about 12 percent of shared variance. Although that may appear to be a small amount of explained variance, one also has to consider practical significance. Being able to better predict that 12 percent of patients would respond favorably to a drug or improving science team innovation by 12 percent based on a leadership or teamwork intervention may be very practically meaningful. Thus, a meta-analysis provides a rigorous quantitative summary of a body of empirical research.

tee believes the evidence on teams in other contexts can be translated and applied to improve the effectiveness of science teams and larger groups.

This chapter begins by presenting critical background information—highlighting key considerations for understanding team effectiveness and presenting theoretical models that conceptualize team processes as the primary mechanisms for promoting team effectiveness. The chapter then highlights those team process factors shown to influence team effectiveness (Kozlowski and Bell, 2003, 2013; Ilgen et al., 2005; Kozlowski and Ilgen, 2006; Mathieu et al., 2008), based on well-established research (i.e., meta-analytic findings [see Box 3-1] or systematic streams of empirical research). Next, the discussion turns to interventions that can be used to improve team processes and thereby contribute to team effectiveness; these are discussed in greater detail in subsequent chapters. This is followed by a discussion of how this foundational knowledge can inform team science, a description of models of team science and effectiveness, and a discussion of areas in which further research is needed to address the challenges emerging from the seven features outlined in Chapter 1.

BACKGROUND: KEY CONSIDERATIONS AND THEORETICAL MODELS AND FRAMEWORKS

Key Considerations

One key consideration regarding team effectiveness is that it is inherently multilevel, composed of individual-, team-, and higher-level influences that unfold over time (Kozlowski and Klein, 2000). This means that, at a minimum, three levels of the system need to be conceptually embraced to understand team effectiveness (i.e., within person over time, individuals within team, and between team or contextual effects; Kozlowski, 2012). Broader systems that encompass the organization, multiple teams, or networks are obviously even more complex. Moreover, individual scientists may be part of multiple research projects spread across many unique teams and thus are "partially included" in their teams (Allport, 1932). As noted in Chapter 1, a recent study suggests that scientists' level of participation (i.e., inclusion) in a team is related to team performance, with higher participation related to increased performance (Cummings and Haas, 2012).

A second critical consideration for understanding, managing, and improving team effectiveness is the degree of complexity of the workflow structure of the team task (Steiner, 1972). In simple structures, team members' individual contributions are pooled together or constructed in a fixed serial sequence. For example, in a multidisciplinary team, members trained in different disciplines combine their expertise in an additive way. Complex structures incorporate the integration of knowledge and tasks through collaboration and feedback links, making the quality of team member interaction more important to team effectiveness.

A final key consideration is the dynamic interactions and evolution of the team over time. According to Kozlowski and Klein (2000, p. 55):

> A phenomenon is emergent when it originates in the cognition, affect, behaviors, or other characteristics of individuals, is amplified by their interactions, and manifests as a higher-level, collective phenomenon.

In other words, emergent phenomena arise from interactions and exchange among individuals over time to yield team-level characteristics. Emergent phenomena unfold over time as part of the team development process. Time is also pertinent with respect to how teams themselves evolve. For example, Cash et al. (2003) reported on the evolution of a transdisciplinary group focused on developing improved varieties of wheat and corn. The authors reported that a strictly sequential approach—in which scientists first developed new crops in the laboratory or field and then later handed them over to native farmers—did not lead to widespread use of the new crops. However, when the native farmers were brought into the research at an earlier point in time, as valued participants and partners with the scientists, the group produced new crops that were widely used. Relatedly, teams have different time frames for interaction (i.e., their life cycle or longevity), and this too will alter the emergent dynamics (e.g., Kozlowski et al., 1999; Kozlowski and Klein, 2000; Marks, Mathieu, and Zaccaro, 2001).

Theoretical Models and Frameworks

Most of the research on team effectiveness has been substantially influenced by the input-process-output (IPO) heuristic posed by McGrath (1964). Inputs comprise (a) the collection of individual differences across team members that determine team composition; (b) team design characteristics (e.g., information, resources); and (c) the nature of the problem that is the focus of the team's work activity. Processes comprise the means by which team members' cognition, motivation, affect, and behavior enable (or inhibit) members to combine their resources to meet task demands.

Although team processes are conceptually dynamic, researchers generally assess them at a single point in time. Hence, they are often represented in the research literature by static perceptions or emergent states (Marks, Mathieu, and Zaccaro, 2001). More recently, team processes have been represented by dynamic or sequential patterns of communications (Gorman, Amazeen, and Cooke, 2010) or actions (Kozlowski, in press). In this report, the committee uses the term "team processes" to refer to both dynamic team processes (e.g., communication patterns) and the emergent perceptual states that result from these processes (e.g., cohesion).

Contemporary theories of team effectiveness build on the IPO heuristic but are more explicit regarding its inherent dynamics. For example, Kozlowski et al. (1996, 1999) and Marks, Mathieu, and Zaccaro (2001) emphasized the cyclical and episodic nature of the IPO linkages. Similarly,

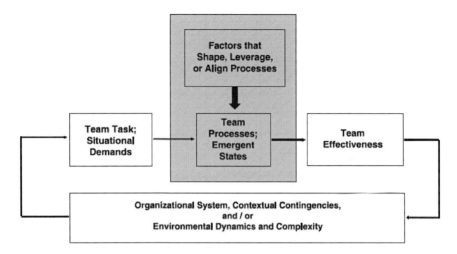

FIGURE 3-1 Theoretical framework and review focus.
SOURCE: Reproduced from Kozlowski and Ilgen (2006). Reprinted with permission.

Ilgen et al. (2005) and Mathieu et al. (2008) are explicit about the feedback loop linking team outputs and subsequent inputs. Accordingly, various authors have urged more attention to team dynamics in research (e.g., Cronin, Weingart, and Todorova, 2011; Cooke et al., 2013) and advances in research design (Kozlowski et al., 2013; Kozlowski, in press) to better capture these dynamics and more clearly specify the relationships between variables. Moving from broad heuristics to more well-defined theoretical models would benefit the field.

In their monograph, Kozlowski and Ilgen (2006) adopted the dynamic IPO conceptualization and focused on those team processes with well-established, empirically supported contributions to team effectiveness. They then considered actions and interventions in three aspects of a team—composition, training, and leadership—that shape team processes and thus can be used to enhance team effectiveness (as shown in the shaded areas of Figure 3-1). Given the preponderance of literature that follows the IPO conceptualization, we emulate that approach in this chapter.

TEAM PROCESSES:
THE UNDERPINNINGS OF TEAM EFFECTIVENESS

Team processes are the means by which team members marshal and co-ordinate their individual resources—cognitive, affective, and behavioral—to meet task demands necessary for collective goal accomplishment. When a team's cognitive, motivational, and behavioral resources are appropriately aligned with task demands, the team is effective. Thus, team processes are the primary leverage point for enhancing team effectiveness. The committee's review in this section examines team cognitive, motivational and affective, and behavioral processes, discussed below.

Cognitive Team Processes

Teams have been characterized as information processing systems (Hinsz, Tindale, and Vollrath, 1997) such that their collective cognition drives task-relevant interactions. Here we discuss several cognitive and perceptual processes that are related to team effectiveness: team mental models and transactive memory, cognitive team interaction, team climate, and psychological safety.

Team Mental Models and Transactive Memory

Team mental models are conceptualized as shared understandings about "task requirements, procedures, and role responsibilities" that guide team performance (Cannon-Bowers, Salas, and Converse, 1993, p. 222). Whereas team mental models represent common understandings, transactive memory captures the distribution of unique knowledge across team members (Wegner, Giuliano, and Hertel, 1985), especially their shared understanding of "who knows what" such that they can access and direct relevant knowledge (Liang, Moreland, and Argote, 1995; Austin, 2003; Lewis, 2003, 2004; Lewis, Lange, and Gillis, 2005; Lewis et al., 2007). Meta-analytic findings indicate that both processes are positively related to team processes ($\rho = .43$) and team performance (i.e., effectiveness) ($\rho = .38$) (DeChurch and Mesmer-Magnus, 2010).

Studies of science teams and larger groups have also found that shared mental models enhance team effectiveness. To cite just a few examples, a study of research and development teams in India (Misra, 2011) found that shared mental models were positively related to team creativity. A study focusing on larger groups of European scientists participating in interdisciplinary and transdisciplinary environmental research found that those groups whose members developed a shared understanding of the research goals were much more likely to succeed in synthesizing their perspectives to achieve those goals than those who did not develop shared understandings (Defila, DiGiulio, and Scheuermann, 2006). In a recent qualitative study of the National Cancer Institute's Transdisciplinary Research on Energetics and Cancer Center, investigators and trainees reported that articulating concrete shared goals (through grant applications, for example) and investing time and effort in developing mutual understanding were essential to successfully carrying out their research projects (Vogel et al., 2014).

Both team mental models and transactive memory have the potential to be shaped in ways that enhance team effectiveness. For example, a number of studies demonstrate that mental models can be influenced by training, leadership, shared or common experiences, and contextual conditions (Cannon-Bowers, 2007; see also Kozlowski and Bell, 2003, 2013; Kozlowski and Ilgen, 2006; Mathieu et al., 2008; Mohammed, Ferzandi, and Hamilton, 2010, for reviews). Similarly, transactive memory systems are formed through shared experiences in working together and training (Bell et al., 2011; see also Blickensderfer, Cannon-Bowers, and Salas, 1997; Kozlowski and Bell, 2003, 2013; Kozlowski and Ilgen, 2006; Mathieu et al., 2008; Mohammed, Ferzandi, and Hamilton, 2010, for reviews). Accordingly, it is often recommended that training be designed to foster development of appropriate team mental models and transactive memory systems and that leaders shape early team developmental experiences to build shared mental models and transactive memory (Kozlowski and Ilgen, 2006).

Cognitive Team Interaction

Team mental models and transactive memory focus on cognitive structure or knowledge and how that knowledge is shared or distributed among team members. Although knowledge certainly contributes to team cognition, it is not equivalent to team-level cognitive processing. Teams often actively engage in cognitive processes, such as decision making, problem solving, situation assessment, planning, and knowledge sharing (Brannick et al., 1995; Letsky et al., 2008). The interdependence of team members necessitates cognitive interaction or coordination, often manifested through communication, the essential building block of team cognition (Cooke et al., 2013). These interactions facilitate information and knowledge sharing processes that are foundational to decision making, problem solving, and the other collaborative cognitive processes mentioned above (Fiore et al., 2010a).

The theory of interactive team cognition proposes that team interaction, often in the form of explicit communication, is at the heart of team cognition and in many cases accounts more than knowledge inputs for variance in team effectiveness (Cooke et al., 2013). In addition, unlike internalized knowledge states, team interaction in the form of communication is readily observable and can be examined over time, thus providing ready access to the temporal dynamics involved (Cooke, Gorman, and Kiekel, 2008; Gorman, Amazeen, and Cooke, 2010).

Another approach to team cognition, focused more on the development of shared problem models, is the macrocognition in teams model (Fiore et al., 2010b). This model is based upon a multidisciplinary theoretical integration that captures the cognitive processes engaged when teams collaboratively solve novel and complex problems. It draws from theories of externalized cognition, team cognition, group communication and problem solving, and collaborative learning (Fiore et al., 2010a). It focuses on team processes supporting movement between internalization and externalization of cognition as teams build knowledge in service of problem solving. Recently the model has been examined in complex contexts such as problem solving for mission control, in which scientists and engineers were required to collaborate to understand and solve problems on the International Space Station (Fiore et al., 2014).

As with other interpersonal processes, interventions can improve cognitive interaction and ultimately team effectiveness. Training that exposes teams to different ways of interacting (Gorman, Cooke, and Amazeen, 2010), as well as team composition changes (Fouse et al., 2011; Gorman and Cooke, 2011), have been found to lead to more adaptive and flexible teams. Similarly, training or professional development designed to support knowledge-building activities has been shown to enhance collaborative problem solving and decision making, leading to improved effectiveness (Rentsch et al., 2010, 2014). These and other professional development approaches are discussed in more detail in Chapter 5.

Science teams and larger groups, like teams in general, are interdependent and require interaction to build new knowledge. They need to manage a range of technological and social factors to coordinate their tasks and goals effectively. Salazar et al. (2012) have proposed a model of team science, discussed later in this chapter, in which social integration processes support cognitive integration processes. These processes can help foster deep knowledge integration in science teams or larger groups.

Many of the features that create challenges for team science described in Chapter 1 introduce challenges to cognitive interaction, and, therefore, interventions that bolster cognitive interaction, such as professional development or training to expose teams to different ways of interacting, may be particularly helpful for science teams.

Team Climate

Climate represents shared perceptions about the strategic imperatives that guide the orientation and actions of team or group members (Schneider and Reichers, 1983; Kozlowski and Hults, 1987). It is always shaped by a particular team or organizational strategy. For example, if a team's goal is to innovate, then the team may have a climate of innovation (Anderson and West, 1998); if the goal is to provide high-quality service, then the team may have a service climate (Schneider, Wheeler, and Cox, 1992); if safety is critical for team or organizational success, then the team or the larger organization may have a safety climate (Zohar, 2000).

Climate has been studied for more than seven decades, and the relationship of climate to important work outcomes is well established (e.g., Carr et al., 2003; Zohar and Hofmann, 2012; Schneider and Barbera, 2013).

Several types of interventions can shape team or group climate. For example, organizations communicate strategic imperatives through policies, practices, and procedures that define the mission, goals, and tasks for teams and larger groups within the organization (James and Jones, 1974). Team leaders shape climate through what they communicate to their teams from higher levels of management and what they emphasize to their team members (Kozlowski and Doherty, 1989; Zohar, 2000, 2002; Zohar and Luria, 2004; Schaubroeck et al., 2012). And team members interact, share their interpretations, and develop shared understandings of what is important in their setting (Rentsch, 1990).

Psychological Safety

Psychological safety is a shared perception among team members indicative of an interpersonal climate that supports risk taking and learning (Edmondson, 1999). The research on psychological safety has been focused primarily on its role in promoting effective error management and learning behaviors in teams (Bell and Kozlowski, 2011; Bell et al., 2011). Learning from errors (i.e., to identify, reflect, and diagnose them and develop appropriate solutions) is particularly important in science as well as in other teams charged with innovation (Edmondson and Nembhard, 2009), and therefore, fostering psychological safety may be uniquely valuable for science teams and larger groups. Although research on this process has not yet been summarized in a published meta-analysis, support for its importance is provided by a systematic stream of theory and research (e.g., Edmondson, 1996, 1999, 2002, 2003; Edmondson, Bohmer, and Pisano, 2001; Edmondson, Dillon, and Roloff, 2007).

Research on psychological safety has focused on the role of team leaders in coaching, reducing power differentials, and fostering inclusion to facilitate psychological safety, so that team members feel comfortable discussing and learning from errors and developing innovative solutions (e.g., Edmondson, Bohmer, and Pisano, 2001; Edmondson, 2003; Nembhard and Edmondson, 2006). Hall et al. (2012a) proposed that creating an environment of psychological safety is critical to lay the groundwork for effective transdisciplinary collaboration. Thus, the research base suggests that appropriate team leadership is a promising way to promote psychological safety, learning, and innovation in science teams and larger groups.

Motivational and Affective Team Processes

Key factors that capture motivational team processes—team cohesion, team efficacy, and team conflict—have well-established relations with team effectiveness.

Team Cohesion

Team cohesion—defined by Festinger (1950, p. 274) to be "the resultant of all the forces acting on the members to remain in the group"—is among the most frequently studied team processes. It is multidimensional, with facets focused on task commitment, social relations, and group pride, although this latter facet has received far less research attention (Beal et al., 2003). Our primary focus is on team task and social cohesion because that is where most of the supporting research is centered.

There have been multiple meta-analyses of team cohesion, with two of the more recent ones (Gully, Devine, and Whitney, 1995; Beal et al., 2003) being the most thorough and rigorous. Both papers concluded that team cohesion is positively related to team effectiveness and that the relationship is moderated by task interdependence such that the cohesion-effectiveness relationship is stronger when team members are more interdependent. For example, Gully et al. (1995) reported that the corrected effect size (ρ) for cohesion and performance was .20 when interdependence was low, but .46 when task interdependence was high. Because high task interdependence is one of the features that creates challenges for team science, fostering cohesion may be particularly valuable for enhancing effectiveness in science teams and larger groups.

Remarkably, although team cohesion has been studied for more than 60 years, very little of the research has focused on antecedents to its development or interventions to foster it. Theory suggests that team composition factors (e.g., personality, demographics; see Chapter 4) and developmental efforts by team leaders (e.g., Kozlowski et al., 1996, 2009) are likely to play an important role in its formation and maintenance.

Team Efficacy

At the individual level, research has established the important contribution of self-efficacy perceptions to goal accomplishment (Stajkovic and Luthans, 1998). Generalized to the team or organizational level, similar, shared perceptions are referred to as team efficacy (Bandura, 1977). Team efficacy influences the difficulty of goals a team sets or accepts, effort directed toward goal accomplishment, and persistence in the face of difficulties and challenges. The contribution of team efficacy to team performance is well established (ρ = .41) (Gully et al., 2002), across a wide variety of team types and work settings (Kozlowski and Ilgen, 2006). As with team cohesion, Gully et al. (2002) reported that team efficacy is more strongly related to team performance when team members are more interdependent (ρ = .09 when interdependence is low, and ρ = .47 when interdependence is high).

Antecedents of team efficacy have not received a great deal of research attention. However, findings about self-efficacy antecedents at the individual level can be extrapolated to the team level. These antecedents include individual differences in goal orientation (i.e., learning, performance, and avoidance orientation; Dweck, 1986; VandeWalle, 1997) and experiences such as enactive mastery, vicarious observation, and verbal persuasion (Bandura, 1977). To develop team efficacy, leaders may consider goal orientation characteristics when selecting team members, but these characteristics can also be primed (i.e., encouraged) by leaders. Similarly, leaders can create mastery experiences, provide opportunities for team members to observe others succeeding, and persuade a team that it is efficacious (see Kozlowski and Ilgen, 2006, for a review).

Team Conflict

Team or group conflict is a multidimensional construct with facets of relationship, task, and process conflict:

> Relationship conflicts involve disagreements among group members about interpersonal issues, such as personality differences or differences in norms and values. Task conflicts entail disagreements among group members about the content and outcomes of the task being performed, whereas process conflicts are disagreements among group members about the logistics of task accomplishment, such as the delegation of tasks and responsibilities (de Wit, Greer, and Jehn, 2012, p. 360).

Although conflict is generally viewed as divisive, early work in this area concluded that although relationship and process conflict were negative factors for team performance, task conflict could be helpful for information sharing and problem solving provided it did not spill over to prompt relationship conflict (e.g., Jehn, 1995, 1997). However, a meta-analysis by De Dreu and Weingart (2003) found that relationship and task conflict were both negatively related to team performance. A more recent meta-analysis (de Wit, Greer, and Jehn, 2012) has shown that the relationships are more nuanced. For example, all three types of conflict had deleterious associations with a variety of group factors including trust, satisfaction, organizational citizenship, and commitment. In addition, relationship and process conflict had negative associations with cohesion and team performance, although the task conflict association with these factors was nil. Thus, this more recent meta-analysis suggests that task conflict may not be a negative factor under some circumstances, but the issue is complex.

Group composition that yields demographic diversity and group fault-lines or fractures is associated with team conflict (Thatcher and Patel, 2011). Because diverse membership is one of the features that creates challenges for team science introduced in Chapter 1, science teams and groups can anticipate the potential for conflict. Many scholars suggest that teams and groups should be prepared to manage conflict when it manifests as a destructive and counterproductive force. Two conflict management strategies can be distinguished (Marks, Mathieu, and Zaccaro, 2001)—reactive (i.e., working through disagreements via problem solving, compromise, and flexibility) or preemptive (i.e., anticipating and guiding conflict in advance via cooperative norms, charters, or other structures to shape conflict processes) (Kozlowski and Bell, 2013).

Team Behavioral Processes

Ultimately, team members have to act to combine their intellectual resources and effort. Researchers have sought to measure the combined behaviors of the team members, or team behavioral processes, in several ways, including by looking at team process competencies and team self-regulation.

Team Process Competencies

One line of research in this area focuses on the underpinnings of good teamwork based on individual competencies (i.e., knowledge and skill) relevant to working well with others. For example, Stevens and Campion (1994) developed a typology of individual teamwork competencies with two primary dimensions (interpersonal knowledge and self-management knowledge) that are each assessed with a set of more specific subdimensions. Based on this typology, they also developed an assessment tool, although empirical evaluations of this tool have yielded somewhat mixed results (Stevens and Campion, 1999).

Others have focused on behavioral processes at the team level. Integrating many years of effort, Marks, Mathieu, and Zaccaro (2001) developed a taxonomy of team behavioral processes focusing on three temporal phases: (1) transition, which involves preparation (e.g., mission, goals, strategy) before task engagement and reflection (e.g., diagnosis, improvement) after; (2) action, which involves active task engagement (e.g., monitoring progress, coordination); and (3) interpersonal processes (e.g., conflict management, motivation), which are viewed as always important.

A recent analysis by LePine and colleagues (2008) extended the Marks,

Mathieu, and Zaccaro (2001) taxonomy to a hierarchical model that conceptualized the discrete behavioral processes as first-order factors loading onto second-order transition, action, and interpersonal factors, which are then loaded onto a third-order, overarching team process factor. Their meta-analytic confirmatory factor analysis found that the first- and second-order processes were positively related to team performance (mostly in the range of $\rho = .25$ to in excess of .30.).

Team Self-Regulation

For teams focused on reasonably well-specified goals, team processes and performance can be related to the team's motivation and self-regulation, similar to models of the relationship between motivation and performance at the individual level. Feelings of individual and team self-efficacy, discussed above (Gully et al., 2002), are jointly part of a multilevel dynamic motivational system of team self-regulation. Team self-regulation affects how team members allocate their resources to perform tasks and adapt as necessary to accomplish goals (DeShon et al., 2004; Chen, Thomas, and Wallace, 2005; Chen et al., 2009). In addition, there is meta-analytic support for the efficacy of group goals for group performance (O'Leary-Kelly, Martocchio, and Frink, 1994; Kleingeld, van Mierlo, and Arends, 2011).

Finally, there is meta-analytic support (Pritchard et al., 2008) for the effectiveness of an intervention designed to increase team regulation by measuring performance and providing structured feedback—the Productivity Measurement and Enhancement System (ProMES; Pritchard et al., 1988). On average and relative to baseline, productivity under ProMES increased 1.16 standard deviations.

Measuring Team Processes

To assess team processes and intervene to improve them, team processes must be measured. Team process factors such as making a contribution to the team's work, keeping the team on track, and appropriately interacting with teammates have traditionally been measured through self or peer reports of team members (Loughry, Ohland, and Moore, 2007; Ohland et al., 2012).

Instruments relying on behavioral observation scales and ratings of trained judges have also been used to measure processes associated with collaborative problem solving and conflict resolution as well as self-management processes such as planning and task coordination (Taggar and Brown, 2001). Brannick et al. (1995) evaluated judges' ratings of processes of assertiveness, decision making/mission analysis, adaptability/flexibility, situation awareness, leadership, and communication. The ratings were found to be psychometrically sound and with reasonable discriminant validity, though the importance of task context was also noted: that is, process needs to be assessed in relation to the ongoing task. "Team dimensional training" was developed to measure a set of core team processes of action teams (e.g., Smith-Jentsch et al., 1998) and has since been validated in numerous settings (e.g., Smith-Jentsch et al., 2008). Another approach that provides for context is the use of checklists of specific processes that are targeted for observation (Fowlkes et al., 1994).

Researchers have measured cognitive processes somewhat differently, relying typically on indirect knowledge elicitation methods such as card sorting to identify team mental models (Mohammed, Klimoski, and Rentsch, 2000) and assess their accuracy (e.g., Smith-Jentsch et al., 2009). In addition, concept maps corresponding to team member mental models have been developed by instructing participants to directly create them (e.g., Marks, Zaccaro, and Mathieu, 2000; Mathieu et al., 2000) or by indirectly creating them through similarity ratings of pairs of concepts analyzed using graphical techniques such as Pathfinder (Schvaneveldt, 1990). Transactive memory systems focusing on team members' knowledge of what each member knows have been measured both via self-assessment (Lewis, 2003) and via communications coding (Hollingshead, 1998; Ellis, 2006). Cooke et al. (2000) reviewed different measurement approaches for measuring team mental models (including process tracing and conceptual methods), pointing out challenges related to knowledge similarity for heterogeneous team members and methods of aggregation.

Recent work in this area has focused on developing measures that are unobtrusive to the teamwork and can capture its complex dynamics (e.g., videorecording, team work simulations, and sociometric badges; Kozlowski, in press). Communication data, for example, can be captured with relatively little interference and provide a continuous record of team interaction (Cooke, Gorman, and Kiekel, 2008; Cooke and Gorman, 2009). This research has identified changes in patterns of simple communication flow (who talks to whom) that are associated with changes in the state of the team (such as loss of situation awareness or conflict). These continuous methods provide a rich view of team process, not captured by static snapshots in time.

INTERVENTIONS THAT SHAPE TEAM
PROCESSES AND EFFECTIVENESS

Table 3-1 identifies actions and interventions that have been found to influence team processes related to three aspects of a team—its composition, professional development, and leadership. This section and the associated three chapters that follow provide detail on each of these three aspects.

TABLE 3-1 Team Processes Related to Team Effectiveness: Interventions and Support

Process	Interventions	Empirical Support for Interventions
Team Mental Models	• Training • Leadership • Shared experience	• Systematic theory, method development, and research • Meta-analytic support (DeChurch and Mesmer-Magnus, 2010)
Transactive Memory	• Face-to-face interaction • Shared experience	• Theory, measurement, and research findings • Meta-analytic support (DeChurch and Mesmer-Magnus, 2010)
Cognitive Team Interaction	• Training • Team composition	• Theory, measurement, and research findings (Gorman, Cooke, and Amazeen, 2010; Gorman and Cooke, 2011)
Team Climate	• Strategic imperatives; team mission/goals; policies, practices, and procedures • Leadership • Team member interaction	• Body of systematic theory, method development, and research (Carr et al., 2003; Zohar and Hofmann, 2012; Schneider and Barbera, 2013)
Psychological Safety	• Leader coaching, inclusion • Positive interpersonal climate	• Systematic empirical support
Team Cohesion	• Antecedents not well specified • Theory = team composition • Theory = leadership	• Systematic empirical support • Meta-analytic support (Gully et al., 1995; Beal et al., 2003)
Team Efficacy	• Mastery experiences • Vicarious observation • Verbal persuasion • Theory = leader behavior	• Systematic empirical support • Meta-analytic support (Gully et al., 2002)
Team Conflict	• Team composition, faultlines • Conflict management skills	• Empirical support • Meta-analytic support (De Dreu and Weingart, 2003; Thatcher and Patal, 2011; de Witt, Greer, and Jehn, 2012)
Team Process Competencies	• Training • Theory = leadership	• Empirical support • Meta-analytic support (LePine et al., 2008)
Team Regulation	• System design • Theory = leadership	• Body of systematic theory and research • Meta-analytic support (Pritchard et al., 2008)

SOURCE: Adapted from Kozlowski and Ilgen (2006). Reprinted with permission.

Team Composition: Individual Inputs to Shape Team Processes

Team composition results from the process of assembling a combination of team members with the expertise, knowledge, and skills necessary for accomplishing team goals and tasks. At the individual level, the logic of staffing is based on selecting individuals with knowledge, skills, abilities, and other characteristics that fit job requirements. At the team level, staffing is more complex because one is composing a combination of members who must collaborate well, not merely matching each person to a well-defined job (Klimoski and Jones, 1995). Chapter 4 takes a detailed look at how team composition and assembly are related to team processes and effectiveness.

Professional Development to Shape Team Processes

Once a team has been assembled, its effectiveness can be facilitated by formal professional development programs (in the research literature, these are referred to as training programs). Although much of the research on team training has focused on programs developed for military teams (Swezey and Salas, 1992; Cannon-Bowers and Salas, 1998), these teams face many of the same process challenges as science teams and groups, resulting from features, such as high diversity of membership, geographic distribution, and deep knowledge integration. Further evidence supporting training as an intervention to facilitate positive team processes is reviewed in Chapter 5, along with discussion of educational programs dedicated to preparing individuals for future participation in team science.

Leadership to Shape Team Processes

Research has shown the influence of leadership on team and organizational effectiveness. Most of this research, however, focuses on the leader, rather than the team, and measures the effectiveness of the leader based on individual perceptions rather than measuring team effectiveness. The leadership literature is also rich with theories of leadership, some of which seem particularly relevant for science teams and larger groups. There is also promising new work on the concept of shared leadership by all team members. Moreover, recent meta-analytic findings provide support for the positive relationship between shared leadership and team effectiveness (42 samples, $\rho = .34$; Wang, Waldman, and Zhang, 2014), suggesting that it may be a useful concept for science teams. Team science leadership is discussed further in Chapter 6.

CONNECTING THE LITERATURE TO TEAM SCIENCE

New Models of Team Science

Researchers have developed and begun to study models of team science and effectiveness. Moving beyond traditional models of group development, such as Tuckman's (1965) phases of storming, norming, forming, and performing, these models incorporate elements specific to science teams and larger groups, such as deep knowledge in interdisciplinary teams, to meet scientific and societal goals. They provide different windows into team science and serve different purposes with respect to team science practice and policy. For instance, Hall et al. (2012b) proposed a model that serves as a heuristic for considering the broad research process. The model delineates four dynamic and recursive phases: development, conceptualization, implementation, and translation (see Box 3-2). Key team and group processes from the literature on teams and organizations are then linked to each of four phases. One of the unique contributions of this model is to highlight the breadth of collaborative and intellectual work that can be done in the early stages of developing a team science research project. Currently, such work in the development phase is often carried out hastily because of resource constraints. This part of the model helps to highlight the need for planning, institutional support, and funding specifically for the development phase. Overall, the model emphasizes key team and larger group processes that may, across the four phases, increase the comprehensiveness and sophistication of the science and effectiveness of the collaboration.

In contrast, Salazar et al. (2012) presented a model that specifically focuses on enhancing a team's integrative capacity through the interplay of social, psychological, and cognitive processes (see Box 3-2). Hadorn and Pohl (2007) presented a model of the transdisciplinary research process that discusses elements of both research and integration processes. The three phases of the model include (1) problem identification and structuring, (2) problem analysis, and (3) bringing results to fruition. This model is specifically designed for incorporating the community perspective (i.e., via "real-world actors") and includes strategies linked to these phases. It draws heavily on a European perspective of transdisciplinarity, science policy, and sustainability research. Reid et al. (2009) and Cash et al. (2003) also discussed models of engaging and integrating knowledge from community stakeholders for sustainability. For instance, Cash et al. (2003) identified key mechanisms for information exchange, transfer, and flow that facilitate communication, translation, and mediation across boundaries in transdisciplinary team science projects.

Existing models of team science have primarily focused on specific aspects of research and knowledge integration processes, but work has

BOX 3-2
Two Models of Team Science

In the first model, Hall et al. (2012b) proposed that transdisciplinary team science includes four phases: development, conceptualization, implementation, and translation:

1. In the development phase, the primary goal is to define the relevant scientific and societal problem. Early in this stage, an informal group of scientists begins to "scope out" a research area and identify relevant areas of expertise. Team and processes critical for effectiveness at this stage include creating a shared mission and goals (i.e., shared mental models); developing critical awareness of the strengths and weaknesses of one's own and other disciplines; and developing an environment of psychological safety. An effective method for supporting these processes is to engage the group in creating a visual representation of the problem area, referred to as a "cognitive artifact," and updating this representation as the work proceeds.

2. In the conceptualization phase, the group develops research questions, hypotheses, a conceptual framework, and a research design. Team processes that enhance effectiveness at this stage include developing shared language, such as by using analogies and lay language in place of disciplinary jargon; developing transactive memory (similar to non-science teams); and developing a transdisciplinary team orientation, which incorporates both the critical awareness described above and team self-efficacy, as described earlier in this chapter.

3. In the implementation phase, the primary goal is to carry out the planned research. The membership of the team or larger group stabilizes as the core participants develop routines, such as frequency of meeting. At this stage, developing a more extensive team or group transactional memory, including shared understanding of how things get done (taskwork) and how interactions occur (teamwork), enhances effectiveness. Conflict management is also essential to avoid conflicts that could otherwise derail the development of team processes. Another critical process at this stage is team learning, including reflection on action, similar to the team regulation approaches described above, while at the same time, scientific effectiveness is enhanced through continued efforts to promote shared language and mental models.

4. In the translation phase, the primary goal is to apply research findings along the research continuum to address real-world problems. As the team or group membership evolves accordingly, developing shared understandings of team goals and roles (i.e., shared mental models and transactive memory) among old and new members aids effectiveness. These processes are especially critical as community or industry stakeholders may become engaged at this stage, potentially creating communication challenges even greater than those involved in communicating across disciplines.

Hall et al. (2012b) proposed that the four-phase model can serve as a roadmap as scientists and stakeholders move through the four phases, and as a guide to evaluation of, and quality improvement for, team science projects.

Salazar et al. (2012) proposed a second model that links the performance of an interdisciplinary or transdisciplinary science team or larger group to its "integrative capacity" defined as the ability to "work across disciplinary, professional, and organizational divides to generate new knowledge . . . through the continuous interplay of social, psychological, and cognitive processes within a team" (Salazar et al., 2012, p. 22).

The authors proposed that integrative capacity allows a team or larger group to overcome barriers to integration that may arise because of several factors, such as team members' strong identification with their individual disciplines, differing conceptualizations of the team goal and the research problem, and geographic dispersion. Thus, the model directly addresses the challenges emerging from several of the key features including high diversity of membership, deep knowledge integration, and geographic dispersion.

The authors identified three pathways that comprise a team's integrative capacity:

1. First, social integration processes, including the development of shared understandings of the project goal (i.e., shared mental models); communication practices facilitated by shared leadership; and collective understanding of all team members' perspectives and expertise (i.e., transactive memory) are the basis for cognitive integration.
2. Second, these social processes lead to emergent states such as trust and positive emotions, which in turn facilitate cognitive integration. Formal interventions, norms, and technological infrastructure can support development of these social processes and emergent states. For example, structured interventions can be used to encourage team members to ask one another about their expertise, supporting development of transactive memory.
3. Third, these social processes and emergent states facilitate the cognitive processes of knowledge consideration, assimilation, and accommodation, leading, in turn, to continued growth of the team's integrative capacity. Feelings of identity with the interdisciplinary science team encourage each team member to thoughtfully consider other team members' knowledge and to either assimilate the new knowledge into his or her own thinking or accommodate it to develop new ways of thinking. Both assimilation and accommodation require reflexivity, or team members' ability to reflect on and improve their own and the team's knowledge, strategy, and processes. Reflexivity is similar to the process of team self-regulation discussed earlier in this chapter, which has been shown to help teams adapt performance as necessary to carry out tasks and accomplish goals.

Although further research is needed to test these two new models of team science, they begin to illuminate how science team processes are related to scientific and translational effectiveness. They also help to address the challenges for team science created by the seven features introduced in Chapter 1.

recently begun on a team science systems map project that would provide a broader, holistic understanding of the system of factors involved in the context, processes, and outcomes of team science (Hall et al., 2014a). Such a map would aid in identifying possible leverage points for interventions to maximize effectiveness, as well as areas where further research is needed.

Features That Create Challenges for Team Science and Team Processes

Most of the key features that create challenges for science teams and larger groups have direct impacts on team processes:

- As noted by Hall et al. (2012b) and Salazar et al. (2012), science teams or larger groups with high diversity of membership (feature #1) face challenges particularly in the area of team process. Communication across scientific disciplines or university boundaries, for instance, may prove difficult.
- Deep knowledge integration (feature #2) is required to achieve the objectives of interdisciplinary or transdisciplinary team science projects, yet also points to team process as a central mechanism for effectiveness. Strategies and interventions to foster positive team processes (described more fully in Chapters 4, 5, and 6) are critical for effective collaboration within science teams and larger groups that have diverse membership and seek to foster deep knowledge integration.
- The research on how team process influences effectiveness described in this chapter has primarily been based on relatively small teams of 10 or less, as few researchers have attempted to conduct empirical team research on larger groups (feature #3). As noted in Chapter 1, most science teams include 10 or fewer members, suggesting that the findings in this chapter are relevant to science teams. Although it is unclear whether the findings scale to larger groups, the committee assumes that increasing size poses a challenge to group processes and ultimately group effectiveness.
- Large science groups composed of subteams that may be misaligned with other subteams (feature #4), as well as teams or groups of any size with permeable boundaries (feature #5), may also be less cohesive than other teams or groups. When team or group membership changes to meet the changing goals of different phases of a transdisciplinary research project, leaders need to make renewed efforts to develop shared understandings of the project goals and individual roles (Hall et al., 2012b). Such efforts, along

with other leadership strategies described in Chapter 6, can help to address these features.

- Geographic dispersion (feature #6) limits face-to-face interaction and development of transactive memory and thereby places a toll on cognitive interaction in a team or group. Some ways to address this particular challenge are described in Chapter 7.

- High task interdependence (feature #7) is often exaggerated in science teams or groups because of the complex demands of scientific research that may involve sharing highly sophisticated technology or carrying out tasks with experts from a different discipline. Increasing task interdependence creates increasing demand for such team processes as shared mental models (shared understanding of research goals and member roles) and transactive memory (knowledge of each team members' expertise relevant to the research goals).

The seven features create challenges through the processes in which science teams engage. The features of diversity, large size, permeable boundaries, and geographic dispersion push team or group members apart, impacting cohesion and conflict and generally challenging cognitive interaction. On the other hand, features such as the need for deep knowledge integration in interdisciplinary and transdisciplinary team or groups and high task interdependence demand enhanced team processes. Thus these features demand high-quality team processes while also posing barriers that thwart them, creating a team process tension.

SUMMARY AND CONCLUSION

Based on its review of the robust research on teams in contexts outside of science and the emerging research on team science, the committee concludes that team processes (such as shared understanding of goals and team member roles, team cohesion, and conflict) are related to effectiveness in science teams and larger groups, and that these processes can be influenced. The committee assumes that research-based actions and interventions developed to positively influence these processes and thereby increase effectiveness in contexts outside of science can be extended and translated to similarly increase the effectiveness of science teams and larger groups. Actions and interventions targeting team composition, team leadership, and team professional development are discussed further in the following chapters.

CONCLUSION. *A strong body of research conducted over several decades has demonstrated that team processes (e.g., shared understanding of team goals and member roles, conflict) are related to team effectiveness. Actions and interventions that foster positive team processes offer the most promising route to enhance team effectiveness; they target three aspects of a team: team composition (assembling the right individuals), team professional development, and team leadership.*

4

Team Composition and Assembly

Together, team composition and assembly make up one of the aspects of a team identified in Chapter 3 that can be manipulated to support team science. Team composition and assembly involve putting together the right set of individuals with relevant expertise to accomplish the team goals and tasks and to maximize team effectiveness.

The first section of the chapter discusses research on team and group composition that can be used to inform strategies for optimizing composition and enhancing effectiveness. Much of this research focuses on how individual characteristics of team or group members are related to performance. However, team composition is more complex than staffing individual positions because the members must collaborate well if the team is to be effective (Klimoski and Jones, 1995). This line of research provides robust evidence based on meta-analyses of empirical work on teams. The second section of the chapter reviews an emerging strand of research—team assembly—that takes a broader focus, examining how both individual characteristics and team processes (including the process of assembling the team or group) are related to team effectiveness. The third section of the chapter discusses tools and methods to facilitate composition and assembly of science teams and larger groups. The fourth section discusses the role of team composition and assembly in addressing the seven features that create challenges for team science outlined in Chapter 1. The chapter ends with conclusions and a recommendation.

TEAM COMPOSITION

Researchers have found that various individual characteristics are important considerations when composing teams or larger groups, both in science and in other contexts. Perhaps the most important individual characteristic to consider when composing a team science project is scientific, technical, or stakeholder expertise. As discussed in previous chapters, one of the key features that creates challenges for team science is high diversity of membership, as it may be necessary to include experts from multiple disciplines and professions to accomplish scientific or translational goals. For example, macrosystems ecology addresses ecological questions and environmental problems at the scale of regions to continents, linking these broad scales to local scales across space and time (Heffernan et al., 2014). Research in this field demands diverse expertise, including information scientists as well as ecologists (Heffernan et al., 2014).

One recent study provides evidence that high diversity of disciplines can improve scientific outcomes: Stvilia et al. (2010) studied 1,415 experiments that were conducted by teams at the national High Magnetic Field laboratory from 2005 to 2008. The authors' analysis of internal documents found that increased disciplinary diversity of the experimental teams was related to increased research productivity, as measured by publications.

Another study, however, illuminates the challenges as well as the benefits of highly diverse membership. In a longitudinal study of more than 500 National Science Foundation-funded research groups, Cummings et al. (2013) found that as the size of research groups increased, research productivity as measured by publications also increased. However, the marginal productivity of the larger groups declined as they became more heterogeneous, either by including experts from more disciplines or from more institutions (Cummings et al., 2013).

Other individual characteristics, including personality traits, may influence team science effectiveness. Feist (2011) has found that the characteristics of eminent, highly creative scientists include not only openness to experience and flexible thinking, but also dominance, arrogance, hostility, and introversion—personality traits that are not associated with being a good team player. Several studies have found that higher intelligence among team members, as measured by a team's mean level of general cognitive ability, is positively related to goal achievement, and the effect sizes are fairly large (e.g., $\rho = .29$ in Devine and Philips, 2001; $\rho = .40$ in Stewart, 2006). Higher conscientiousness, measured as a team's mean conscientiousness, is also positively related to team performance, although the relationship is stronger for performance and planning tasks than it is for creative and decision-making tasks that are similar to those carried out by science teams (Koslowski and Bell, 2003).

Extroverts who can easily monitor and respond appropriately to actions and attitudes of others (McCrae and Costa, 1999) may work more effectively in science teams or larger groups than introverts who are less attuned to teammates' actions and attitudes (Olson and Olson, 2014). Some evidence supports this theory, indicating that teams with higher mean levels of extroversion are more effective than teams with lower levels of this personality trait[1] (Kozlowski and Bell, 2003). Woolley et al. (2010) recently identified a new individual construct related to extroversion that they refer to as "social sensitivity," as well as a team-level construct called "collective intelligence." In two studies of nearly 700 people working in small groups, the authors found evidence of the team-level collective intelligence factor and showed that it was related to group performance on a variety of tasks. The new factor was not strongly correlated with the mean level of intelligence within a group, but it was significantly correlated with the mean level of social sensitivity, the level of equality in taking turns during group discussions, and the proportion of females in the group. Individual social sensitivity was measured using a test requiring participants to "read" the mental states of others from looking at their eyes. In a follow-up study focusing on online groups, Engel et al. (2014) again found that a group's level of general collective intelligence was related to performance across a variety of tasks and that social sensitivity of group members was significantly related to collective intelligence. The result was surprising because social sensitivity was measured using the same test of one's ability to discern another's mental states by looking at their eyes and face, although the members of the online groups never saw each other at all. This suggests that the test measures a deeper aspect of an individual's ability to discern the mental states of others, beyond what the individual can "read" from another's eyes and facial expressions.

Based on data collected using unobtrusive badges to record team member interactions, Pentland (2012) also found that the level of equality in taking turns when speaking was related to team performance. He proposed that the most valuable team members are "charismatic connectors," who circulate among all team members and spend equal amounts of time listening and speaking, while also seeking ideas outside the team; in a study of business leaders attending an executive education program, he found that the more charismatic connectors were included in a team, the more successful the team was. Finally, another related construct—the disposition to forge connections and share information among groups and individuals—

---

[1] When considering potential members for a team or larger group, it is important to recognize that individuals lacking in a beneficial characteristic (e.g., social or communication skills related to extroversion) may develop it through education or professional development, as discussed in Chapter 5.

was studied in the engineering division of an auto manufacturing firm. Individuals with this disposition were more frequently involved in innovation than other individuals (Obstfeld, 2005).

Although the finding that a high level of general cognitive ability enhances team effectiveness might suggest that science teams and groups should be composed entirely of individuals with this characteristic, a balance of characteristics may most benefit team effectiveness (Kozlowski and Ilgen, 2006). For example, a team composed entirely of extroverts might focus more on socializing than completing tasks while a team of highly conscientious individuals might be so task focused that the members do not collaborate well. Little research has tested this theory; however, one study of 41 teams in a research and development firm used an assessment to assign the team members into one of three cognitive styles: creativity, conformity to rules, and attention to detail (Miron-spektor, Erez, and Naveh, 2011). The authors found that including a balance of both creative and conformist members on a team enhanced its radical innovation (characterized as developing something completely new), whereas including a higher proportion of attentive-to-detail members hindered radical innovation (Miron-spektor et al., 2011). More recently, Swaab et al. (2014) found that basketball and soccer teams (which require highly interdependent actions by teammates) with the highest proportion of the most talented athletes performed worse than teams with more moderate proportions of the most talented athletes.

Other individual differences on dimensions such as gender, ethnicity, age, and specialized knowledge and abilities have been shown to exert both positive and negative influences on group processes and effectiveness. However, it is important to note that, in general, these other individual differences show smaller effects than do those discussed above (average level of cognitive ability, conscientiousness) (Bell et al., 2011). Prior studies that have examined the influence of individual differences and team diversity on team functioning generally have focused on one characteristic (or very few) at a time. However, each individual brings multiple characteristics to the team, making it difficult to prescribe individual factors for ideal team composition. By contrast, an emerging line of research on group faultlines (defined and discussed further below) takes into account the interplay among diverse individual characteristics and has made substantial progress in the past decade (Lau and Murningham, 1998; Chao and Moon, 2005; Thatcher and Patel, 2011; Carton and Cummings, 2012, 2013; Mathieu et al., 2014).

Here we highlight general findings for team composition based on team diversity, group faultlines, team subgroups, and changing team membership—factors that have clear implications for team science effectiveness.

Team Diversity

Diversity is at the heart of being a team, as teams have been defined as groups of individuals with different roles who work interdependently (Swezey and Salas, 1992). Indeed, interdisciplinary science teams and groups can be characterized this way (Fiore, 2008), making diversity the rule, not the exception. Research in this area has generally been conducted under the theoretical assumption that greater heterogeneity is associated with more diverse perspectives and, hence, better quality outcomes for diverse groups (Jackson, May, and Whitney, 1995; Mannix and Neale, 2005). However, support for this optimistic view has proven to be elusive and mixed at best, with findings supporting positive (Gladstein, 1984), negative (Wiersema and Bird, 1993), and no relationships (Campion, Medsker, and Higgs, 1993).

In their narrative review, Mannix and Neale (2005) concluded that demographic heterogeneity (based on easily recognizable surface features of an individual, such as gender, race, or age) tends to impede the ability of group members to collaborate effectively, whereas heterogeneity of knowledge and personality types—that is more task relevant—is more often associated with positive outcomes, but only when group processes are appropriately aligned with the task. A meta-analysis of the team diversity literature by Horwitz and Horwitz (2007) found no relationship between demographic diversity and the quality and quantity of team outcomes and small but statistically significant positive relationships between task-related diversity and the quality ($\rho = .13$) and quantity ($\rho = .07$) of team outcomes. A subsequent and larger meta-analysis by Joshi and Roh (2009) examined how contextual factors influenced the relationship between task-related diversity, demographic diversity, and team effectiveness. The authors found that contextual factors such as team interdependence and occupational setting influenced the direction and level of the relationships. For example, gender diversity had a significant negative effect on team performance in male-dominated occupational settings but a significant positive effect on team performance in gender-balanced occupational settings.

In light of the small and mixed effect sizes in previous studies of the relationship between diversity and team performance, Bell et al. (2011) conducted a new meta-analysis. The authors distinguished between the various conceptualizations of "diversity" used in previous studies, including diversity variety (multiple sources of expertise or knowledge that may contribute to team effectiveness), diversity separation (similarities or differences among team members that may lead to subgroups and negatively affect performance), and disparity (inequality within the team, such as the inclusion of one very senior member and many newcomers that may affect performance). They examined specific variables, rather than clusters of "job-related" (i.e., task-related) and "demographic" or "less job-related" variables, and considered how different performance measures and team types influenced the relationship. Significantly for team science, the performance measures included innovation or creativity, as well as general performance, and the team types included design teams charged with creating and designing new products.[2] The authors found that only one type of task-related diversity—functional background diversity (i.e., the organizational division or profession of the team members)—had a small positive relationship with general team performance ($\rho = .11$). This relationship was larger when the performance measure was innovation or creativity ($\rho = .18$) and for design teams compared with teams in general ($\rho = .16$). In contrast, race variety diversity and gender diversity were negatively related to team performance ($\rho = -.13$ and $-.09$, respectively). Age diversity was unrelated to team performance.

In contrast to these meta-analytic findings, two recent studies focusing specifically on science found positive relationships between demographic and national diversity and the effectiveness of science teams or groups. First, Freeman and Huang (2014) studied the citation rates of more than 1.5 million scientific papers, finding that persons of similar ethnicity co-author together more frequently than can be explained by chance given their proportions in the population of authors and that this homogeneity in authoring teams or groups is associated with weaker scientific contributions (as measured by citations). Papers produced by authors of diverse ethnicities are cited more frequently than those produced by authors of similar ethnicity. Freeman and Huang (2014a) proposed that ethnic diversity reflects idea diversity and, thus, better science is produced when collaborators bring different ideas and ways of thinking to the effort. They found the same positive effect on citations when researchers from geographically diverse universities collaborated. In a further analysis including 2.5 million papers, Freeman and Huang (2014b) again found that papers produced by authors of diverse ethnicities are cited more frequently than those produced by authors of similar ethnicity. Second, Smith et al. (2014) analyzed all papers published between 1996 and 2012 in eight disciplines, finding that those with more countries in their affiliations performed better in journal placement and citation performance than those whose authors came from fewer countries.

Other studies suggest that gender diversity can be beneficial for team science, showing that women tend to collaborate more than men do in academic science (Bozeman and Gaughan, 2011; Rijnsoever and Hessles, 2011). As noted above, Woolley et al. (2010) found that the proportion

---

[2] As discussed in Chapter 1, new product development teams experience many of the same challenges as science teams.

of women in a group was related to the group's collective intelligence, or ability to perform a variety of tasks. Bear and Woolley (2011) reported that the presence of women on teams is associated with improved collaborative processes. These processes have been shown to increase team effectiveness, as discussed in Chapter 3.

Overall, the research findings on the facilitative or inhibiting aspects of team diversity are mixed, although the meta-analytic evidence clarifies the picture somewhat. Further research is needed to explore how various forms of diversity are related to team performance. Following Bell et al. (2011), it will be important to carefully articulate the theoretical connection between the specific variable, the conceptualization of diversity, and team performance.

Group Faultlines

Faultlines are hypothetical divisions within a team based on team composition (e.g., two biologists and two physicists in a team form a possible faultline based on discipline). When compositional differences among members are made salient, such as when the team has to decide how to allocate resources or how to divide up the work, faultlines are said to be "activated" and subgroups are formed, raising potential for conflict (Bezrukova, 2013). For example, if a science team including two biologists and two physicists has enough funding to hire only one doctoral student, then faultlines may be activated as each disciplinary group wants to hire a student within its discipline.

Although the faultline concept is relatively new to the literature, it has stimulated a substantial amount of research, enabling an integrative and informative meta-analytic review by Thatcher and Patel (2011). Essentially, research in this area supports the differential effects of task-relevant and demographic diversity on team effectiveness: demographic diversity (factors such as gender, race, age, and tenure) is related to faultline strength, whereas task-related diversity in factors, such as educational level and experience with a team function are as well, but less so. Faultline strength contributes to weakened team relationships and task conflict that, in turn, inhibit team member satisfaction and performance. However, managers can address this problem by fostering identification with the larger team and developing shared goals (Bezrukova, 2009, 2013; see further discussion in Chapter 6).

Subgroups in Teams

Going beyond the faultline concept, Carton and Cummings (2012, 2013) have developed an alternative conceptualization of subgroup formation. Subgroups are subsets of team members who are uniquely interdependent in some way, such as those members who develop friendships with each other or who choose to collaborate. Prior empirical work has highlighted some of the benefits and costs of subgroup formation in teams. For example, in a study of 156 teams in pharmaceutical and medical products firms, Gibson and Vermeulen (2003) found that subgroup strength (i.e., the extent to which members in a subgroup overlapped on attributes, such as age, gender, ethnicity, function, and tenure) facilitated team learning behaviors. Teams with subgroups who had more in common were better able to come up with new ideas, communicate with each other, and document what they learned. However, when Polzer et al. (2006) examined the impact of subgroups within geographically dispersed teams, they found that teams including subgroups based on geography experienced higher conflict and lower trust. In particular, conflict was highest and trust was lowest when there were two equally sized subgroups each in a different country.

Other research findings have also illustrated the challenges of communicating across subgroups when faultlines are stronger, when subgroup distance is greater (e.g., subgroups based on very different ages; Bezrukova et al., 2009), and when subgroup size is imbalanced (e.g., six members in one subgroup and two members in another subgroup; O'Leary and Mortensen, 2010). A recent study by Carton and Cummings (2013) begins to reconcile some of the different results around the impact of subgroups in teams. They show that having more balanced subgroups can be better for team performance if the subgroups are knowledge-based (e.g., members with the same business unit and reporting channel in the organization) but worse for team performance if the subgroups are based on demographic characteristics, such as the same age and gender. On the one hand, in the case of knowledge-based subgroups, having an equal representation of knowledge sources on the team can be beneficial for integrating what is known (van Knippenberg, De Dreu, and Homan, 2004). On the other hand, having two subgroups composed of members with the same demographic characteristics can be costly when members get locked into in-group/out-group differences (Tajfel and Turner, 1986).

Recent research provides insights on how to manage subgroups, whether based on knowledge or demographic characteristics. For example, Sonnenwald (2007) discussed some of the issues that can arise, such as mistrust, misunderstanding, and conflict, when ethnic minorities and minority-serving institutions participate in team science. He reported on strategies to address these issues, which include conducting extensive outreach to all participants early in the research planning, convening facilitated discussions with community authorities (e.g., religious leaders, tribal leaders), and using focus groups to elicit the community's concerns and priorities related to the research. DeChurch and Zaccaro (2013) identified leadership strategies to mitigate competition between teams within a larger multi-team system (similar to subgroups within a team) and foster shared identification with high-level goals (see Chapter 6 for further discussion). Structured discussions can be used to foster communication across subgroups based on discipline (O'Rourke and Crowley, 2013; see Chapter 5).

Changing Team Membership

Recent empirical work on teams, though not supported by meta-analytic findings, nonetheless suggests that changing team membership can enhance team performance. Gorman and Cooke (2011) found that in a three-person military command and control task, changing team members in a second session resulted in teams that were more adaptive in that they could better respond to novel events. In another study (Fouse et al., 2011), it was found that simply changing the location of team members doing a military planning task around a table resulted in a superior plan score compared to teams whose members stayed in the same location. Gorman and Cooke (2011) hypothesized that changes in team membership provide a chance for team members to experience more diversity in process behaviors, which is useful when the team faces challenges requiring different approaches. Similarly, changes in group membership associated with members leaving a group for another and then returning have been associated with increased creative ideas in essay writing (Gruenfeld, Martorana, and Fan, 2000). There seems to be some evidence for the positive influence of changing team membership from studies conducted outside of the laboratory. Kahn (1993) described the value of adjusting the composition of interdisciplinary science teams over the life cycle of a research network supported by the MacArthur Foundation.

Changing team composition through membership changes, often considered detrimental to team effectiveness, seems in some instances to have a positive effect and might be a useful intervention. In particular, faultlines that have formed may be disrupted by changing membership and collaboration dynamics that may be dysfunctional to team effectiveness may be pushed off their trajectory, resulting in positive process change.

TEAM ASSEMBLY

Science teams and larger groups may be assembled by individual scientists, university research administrators (who sometimes function as matchmakers; see Murphy, 2013), funding agencies, or other groups or individuals. To guide the assembly process, individuals or organizations may rely on information about potential teammates based on prior relationships, consultations with experts in relevant areas, or more structured information sources. A new strand of research, known as team assembly, examines not only the composition of the team but also these processes.

Research on team assembly examines team composition at the team level (including the fit between team and task), the relational level within the team (e.g., individuals' prior relationships with each other), and the ecosystem surrounding the team (National Research Council, 2013). The goal is to understand how these multiple levels influence team performance. Here, we briefly discuss some of the findings from this new strand of research.

Guimera et al. (2005) studied science team formation, composition, and performance based on the analysis of teams in another domain—the universe of creative artist teams that made Broadway musicals from 1950 to 1995. Both Broadway and scientific teams aim to advance novel ideas and be creative (Uzzi et al., 2013). The authors found that Broadway teams were composed of two fundamental types of teammates: newcomers and experienced incumbents. They then defined the relationships within the team as newcomer-newcomer, newcomer-incumbent, incumbent-incumbent, and incumbent-repeated ties, finding that musical teams including a mix of all four types of relationships were most successful.

Guimera et al. (2005) applied this framework to science teams in four academic disciplines: astronomy, ecology, economics, and social psychology. Data on team composition were derived from authorship data from the five to seven top journals in each field, circa 1955–2004, as recorded in the Web of Science. They found that science team performance, as measured by the average citations accumulated by a paper (i.e., the journal impact factor), was positively associated with the probability of incumbents on the team, but only if the team had diversity, including newcomers and repeated ties among incumbents on the team. It is important to note that the model is predictive first and foremost of the population's performance level, not individual team-level performance. Consequently, any one team can be an exception in the short run, whereas the long-run systemic network within which teams in a field are embedded predicts average team performance in that field.

Contractor et al. (2014) conducted a study of student teams focusing on how they were assembled. First, students could either be assigned to teams or they could self-organize. Second, students could either use unstructured information about the other individuals to select teammates or use a team-builder tool populated with data provided by students with information about their attributes, social networks, and the sorts of people they would like on their team. The researchers found that teams that had used the team-builder tool were more homogeneous in age and cultural sensitivity, but more heterogeneous by sex. Not surprisingly, the self-organized teams (whether or not they used the builder) were more likely to contain members who had previously worked together than the teams that were assigned randomly. Analysis of surveys conducted 4 weeks after team formation showed that teams whose members all played a role in their organization (whether by using the builder or simply choosing their friends) communicated more and were more confident in their ability to work together effectively than teams with any members who were assigned.

Findings such as these raise questions about funding requirements that mandate inclusion of certain individuals, scientific disciplines, or institutions, within a team or larger group, rather than allowing teams or groups to self-organize. On the other hand, self-organizing teams or groups may be composed primarily of individuals with prior collaborative relationships, missing the benefits of newcomers with innovative ideas.

METHODS FOR FACILITATING COMPOSITION AND ASSEMBLY OF SCIENCE TEAMS AND GROUPS

When the general focus of a research and/or translational problem has been established, team assembly can be guided using a "person-task fit" approach, or matching characteristics of individuals with characteristics of the research and/or translational task (National Research Council, 2013). Fields such as human factors (Wickens et al., 1997) and cognitive engineering (Lee and Kirlik, 2013) have contributed a number of methods for analyzing tasks that can guide team assembly. Task analysis involves the systematic decomposition of the behavior required of a task in order to understand the human performance requirements (Kirwan and Ainsworth, 1992). When composing a science team or group, it may be important to understand the tasks involved in operating scientific tools or equipment that will likely require specific technical competencies of one or more team members.

Assembly of science teams and groups may also benefit from cognitive engineering methods. Cognitive architectures, such as ACT-R, social network models, and agent-based modeling, have been used to understand and improve team effectiveness in highly cognitive tasks and can also be used to guide team assembly (Kozlowski and Ilgen, 2006). In addition, task analytic methods such as Cognitive Work Analysis (Vincente, 1999) have been used to design teams for first-of-a-kind work systems (Naikar et al., 2003). The fact that these complex systems are first of a kind makes the early analysis challenging, but in essence, the task model is developed alongside requirements for the team. This method takes advantages of constraints in the work environment that influence behavior. It involves detailed observations of work in context, accompanied by interviews at various levels of the organizational hierarchy to develop an understanding of the task or work in context. This approach has been applied to complex sociotechnical systems in which there are many people working with complex technology. Some science teams and groups work in similar environments, where they collaborate in designing and operating large and complex scientific equipment that is shared (e.g., the Large Hadron Collider). There are no data on the effectiveness of teams designed using this approach; however, it provides an analytic way of decomposing a task and work environment that may suggest team design needs that would otherwise be missed. These cognitive engineering approaches provide a systematic way of determining team requirements in terms of knowledge, skills, and abilities that can be used to guide team composition and assembly.

In other cases, however, the problems to be addressed using a team science approach are not clearly defined. As noted in the previous chapter, a team science project may begin when a group of scientists and/or stakeholders comes together to explore a problem or issue and the first phases may involve clarifying the focus and delineating research questions (Hall et al., 2012b; Huutoniemi and Tapio, 2014). In these cases, information on the larger ecosystem—the network of scientists and stakeholders with relevant interests and knowledge—may be helpful for team assembly.

Surveys have found that scientists, university administrators, and others involved in assembling science teams need a variety of information about potential collaborators, including not only publications, but also research interests, grant topics, and patents (Obeid et al., 2014). Such information is available from research networking systems that use data mining and social network approaches to create large, easily searchable databases, facilitating the search for scientific collaborators. These systems enable users to discover research expertise across multiple disciplines; identify potential collaborators, mentors, or expert reviewers; and assemble science teams based on publication history, grants, and/or biographical data (Obeid et al., 2014).

Many research networking tools are available, including Biomed Experts;[3] Elsevier's SciVal© Experts and Pure Experts Portal;[4] Harvard Catalyst Profiles;[5] DIRECT: Distributed Interoperable Research Experts Collaboration Tool;[6] and VIVO (Börner et al., 2012). VIVO, for example, is a free, open-source web application developed with support from the National Institutes of Health that facilitates search of researchers by publications, research, teaching, and professional affiliations across institutional

---

[3] Biomed Experts, see http://www.biomedexperts.com [April 2015].

[4] Elsevier's SciVal© Experts and Pure Experts Portal, see http://www.elsevier.com/online-tools/research-intelligence/products-and-services/pure [April 2015].

[5] Harvard Catalyst Profiles, see https://connects.catalyst.harvard.edu/profiles/search/ [May 2015].

[6] DIRECT: Distributed Interoperable Research Experts Collaboration Tool, see http://direct-2experts.org/?pg=home [April 2015].

boundaries (Börner et al., 2012). My Dream Team Assembler builds upon VIVO to incorporate social network analysis and modeling of the seeker to make recommendations of potential scientific collaborators (Contractor, 2013). An evaluation guide[7] to research networking systems is available to assist institutions as they consider adopting these new tools.

Recent surveys suggest that research universities, especially academic medical centers, are increasingly adopting research networking systems (Murphy et al., 2012; Obeid et al., 2014), and many plan to share data on research expertise at their institutions using linked open data, allowing it to be widely accessed and analyzed. These publicly available data show promise for use in assessing cross-institution research collaborations in future team science research (Obeid et al., 2014). A recent study of implementation at the University of California at San Francisco (Kahlon et al., 2014) found that the research networking system was attracting an increasingly large pool of visitors whose behavior suggested they were using the tool to identify new collaborators or research topics. In response to an online survey, users identified a range of benefits to using the system to support research and clinical work. With the exception of this one study, however, there is little evidence to date that using the tools to guide team assembly results in teams or groups that are more effective than other teams or groups. The committee suggests that practitioners who choose to try one or more of these tools track the tools' usefulness and usability in assembling teams and collaborate with researchers to assess their impact on scientific outcomes.

ADDRESSING THE SEVEN FEATURES THAT CREATE CHALLENGES FOR TEAM SCIENCE

How does the research on team composition and assembly speak to each of the seven features that create challenges for team science?

High diversity of membership (feature #1) is directly addressed by the research in team composition, faultlines, and subgroups summarized above. The finding that task-related diversity is associated with more effective teams is a promising finding for team science projects, which are composed primarily on the basis of task diversity.

Deep knowledge integration (feature #2) is actually a result of team composition, given that team science projects often require the integration of knowledge from multiple disciplines and stakeholders. Some of the tools discussed above such as the research networking systems, can potentially help mitigate the communication challenges resulting from this feature by

---

[7]Evaluation Guide, see https://www.teamsciencetoolkit.cancer.gov/public/TSResourceTool. aspx?tid=1&rid=743 [April 2015].

making it possible to learn more about potential teammates in advance of team or group formation.

Large size (feature #3) is moderated by the heterogeneity of team or group members such that larger groups have been found to be more productive, but this advantage over smaller teams declines with increased heterogeneity in the disciplines and institutions represented (Cummings et al., 2013). Using methods such as cognitive work analysis to carefully analyze the tasks and requirements for team or group members of varying disciplines would help avoid unnecessary challenges of size and diversity.

The challenges emerging from goal misalignment with other teams (feature #4) are consistent with the concept of faultlines and subgroups that can be avoided by careful attention to team or group composition. However, science leaders or funding agencies sometimes place additional constraints on composition by requiring that a team or group include certain types of individuals, scientific disciplines, or institutions. Such constraints can inadvertently bring together subteams with multiple and sometimes conflicting goals. In these cases, it may be difficult to avoid the development of subgroups, and leadership and professional development interventions can be directed toward increasing the alignment of all subgroups with the high-level goals of the larger group.

Permeable team and group boundaries (feature #5) have been addressed only recently by research on dynamic team membership that acknowledges that modern teams tend to have fluid boundaries (Mathieu et al., 2014). Tannenbaum et al. (2012) observed as well that because organizations often need to rapidly reconfigure teams, individuals increasingly participate simultaneously in multiple teams. They noted that membership fluidity has been found to have both positive and negative effects on team performance, facilitating knowledge transfer on one hand, yet potentially reducing team members' bonds of affiliation on the other hand. To address these challenges, the authors suggested using team assembly tools, increasing role clarity, developing transportable team competencies, and focusing on team handoffs and transitions. At the same time, team processes may in fact be strengthened by changes in team membership as a result of increased team flexibility and adaptivity (Gorman and Cooke, 2011), increased unique ideas (Gruenfeld, Martorana, and Fan, 2000), and improved transfer of knowledge and alignment of member knowledge, skills, and abilities with task demands (Tannenbaum et al., 2012). Some research has found that acquaintance among team members and the trust it engenders facilitates effectiveness in cross-institutional teams or groups (Gulati, 1995; Shrum, Genuth, and Chompalov, 2007; Cummings and Kiesler, 2008). But, as discussed earlier, other studies suggest that membership changes and inclusion of members who are not prior acquaintances can improve the effectiveness of science teams or larger groups (Pelz and Andrews, 1976; Kahn and Prager, 1994; Guimera et al., 2005).

Geographic dispersion (feature #6) is known to create challenges for team success. Polzer et al. (2006) found that having subgroups based on geography was associated with higher conflict and lower trust. Geographically dispersed science team or groups are more likely to be successful if they are assembled so as to avoid faultlines and subgroups known to be problematic. However, if the scientific problem demands inclusion of members who may potentially divide along faultlines, interventions such as those described in Chapter 7 may be warranted.

Finally, high task interdependence (feature #7), a feature of many science teams and larger groups, can generate challenges when interdependence is required across subgroups or faultlines. Balancing teams at assembly to avoid such faultlines or counteracting them via leadership or other interventions will help facilitate interdependent work.

SUMMARY, CONCLUSIONS, AND RECOMMENDATION

Most of the studies of the relationship between team composition and team effectiveness have yielded conflicting or weak effects. However, task-relevant heterogeneity does seem to be related to team effectiveness with important implications for science teams or groups including multiple disciplines. Further research on faultlines and the subgroups that can result from them corroborate the positive influence of task-related heterogeneity and the need to carefully manage demographic heterogeneity. At the same time, emerging research suggests that demographic heterogeneity can sometimes support scientific productivity.

The recent research on team assembly is beginning to offer insights into how the process of assembling the team or group and the prior relationships between the members affects the scientific and translational outcomes of team science. Research networking systems show promise for helping individual scientists, university research administrators, funders, and others identify potential team members. Further research on team assembly would be valuable at a time of rapid growth in team science.

The committee views this body of work as preliminary evidence that team composition and assembly matter and require careful management to facilitate effectiveness (Fiore, 2008). It is important to recognize that assembling and composing the team provides the raw building material for an effective team and therefore is a critical step, but it is only the first step toward an effective group or team (Hackman, 2012). Ployhart and Moliterno (2011) pointed out that human capital originates in the knowledge, skills, abilities, and other characteristics of individuals, but is transformed into a team resource through interpersonal processes such as those described in Chapter 3. Interventions in other aspects of a teams or groups, beyond composition and assembly, are important to support positive team processes and effectiveness, and we discuss these other aspects in the following chapters.

CONCLUSION. *Research to date in non-science contexts has found that team composition influences team effectiveness, and this relationship depends on the complexity of the task, the degree of interdependence among team members, and how long the team is together. Task-relevant diversity is critical and has a positive influence on team effectiveness.*

CONCLUSION. *Task analytic methods developed in non-science contexts and research networking tools developed in science contexts allow practitioners to consider team composition systematically.*

RECOMMENDATION 1: Team science leaders and others involved in assembling science teams and larger groups should consider making use of task analytic methods (e.g., task analysis, cognitive modeling, job analysis, cognitive work analysis) and tools that help identify the knowledge, skills, and attitudes required for effective performance of the project so that task-related diversity among team or group members can best match project needs. They should also consider applying tools such as research networking systems designed to facilitate assembly of science teams and partner with researchers to evaluate and refine these tools and task analytic methods.

5

Professional Development and Education for Team Science

In Chapter 3, the committee concluded that training interventions offer a promising route to increase team effectiveness. This chapter builds on that conclusion, reviewing research on team training and education for team science. The chapter begins with an introduction to team training, its goals and effectiveness. The second section reviews team-training interventions that show promise for increasing the effectiveness of science teams and larger groups, and the third section reviews interventions designed specifically for team science. The fourth section focuses on education for team science. The fifth section reviews training and education strategies that can help to address the challenges emerging from the seven features introduced in Chapter 1. The chapter ends with a summary, conclusions, and a recommendation.

As a preface to the chapter, we note that professional development, education, and training are general terms that are too often used without clear definitions. The terms "training" or "professional development" can be used to describe a variety of learning activities, ranging from an hour-long presentation on a given scientific topic to a weekend retreat about managing team conflict. The word "education" might be used to describe the same hour-long presentation on a scientific topic or an undergraduate course designed to teach students from different disciplines how to work together on team projects. The context can provide some clues. In universities, the terms "professional development" or "training" are typically used to describe activities outside the classroom, such as research experiences, while the word "education" refers to in-class learning experiences. But, even in academic contexts, confusion can arise. For example, when doctoral students attend an hour-long presentation on a scientific topic related to their research, should the learning experience be called education, professional development, or training? When postdoctoral fellows, who have completed their formal education, attend the same presentation, should it now be called professional development or training?

In sum, the use of the terms "education" and "training" both in the research literature and in practice can sometimes be arbitrary, although which term is used may affect how learning processes and outcomes are measured and funding is allocated. Despite these important distinctions, for sake of reviewing the literature in this chapter, we use the terms adopted by the authors of each study. In future research, it will be important to delineate more clearly the meaning of these teams to develop greater coherence in science policy and practice.

GOALS AND EFFECTIVENESS OF TEAM TRAINING

Generally, team training is defined as an intervention to improve team performance by teaching competencies necessary for effective performance as a team (Cannon-Bowers et al., 1995; Delise, Gorman, and Brooks, 2010). Drawing from the decades-long tradition of learning research in psychology and education, Kraiger, Ford, and Salas (1993) argued for organizing the desired learning outcomes of training in terms of knowledge, skills, and attitudes. The same three categories of learning outcomes have been adopted in the team-training literature, as follows (Cannon-Bowers et al., 1995; for reviews, see Salas et al., 1999; Salas, Cooke, and Rosen, 2008; Klein et al., 2009; Delise, Gorman, and Brooks, 2010; Shuffler, DiazGranados, and Salas, 2011):

- team knowledge (e.g., task understanding, shared mental models, role knowledge)
- team skills (e.g., communication, assertiveness, situation assessment); and
- team attitudes (e.g., team orientation, trust, cohesion).[1]

Training for a particular team is often designed based on analysis of the situational and environmental context, which establishes team goals and tasks and enables identification of the needed knowledge, skills, and attitudes (Bowers, Jentsch, and Salas, 2000).

---

[1] Research on educational preparation for team science has also organized the desired learning outcomes into these same three categories, as discussed later in this chapter (e.g., Nash, 2008).

Recent research provides a more detailed framework of team knowledge, skills, and attitudes (which we refer to as "competencies") emerging from the team context, as well as the situational and environmental context that can be used to design training strategies. First, team training may focus on either *taskwork* or *teamwork* competencies (or both). Taskwork training targets the improvement of task-specific competencies (for science teams and groups, this would include scientific knowledge and skills related to the research problem), while teamwork training targets the improvement of team collaboration competencies. Building on the distinction between taskwork and teamwork proposed by Cannon-Bowers et al. (1995), Fiore and Bedwell (2011) described four types of team competencies for science teams and groups: (1) context-driven competencies specific to a given task and team; (2) team-contingent competencies that are relevant to a particular team but can be applied across various tasks; (3) task-contingent competencies that are relevant to a particular task, regardless of what team performs the task; and (4) transportable competencies, which can be applied across tasks and teams.

Cannon-Bowers et al. (1995) suggested that the first three types of competencies (specific to the task and/or the team) be developed through training for the team as a whole, while the more general "transportable" competencies be developed through education for individuals. Research on training and learning has shown that transfer of training is facilitated when the training context is similar to the context in which the trained skills will be applied (i.e., the workplace). Because the first three types of competencies are specific to a particular task and team context, Cannon-Bowers et al. (1995) suggested that training in these competencies be provided to intact teams (the specific team context) in their real work contexts or simulations of these contexts. Similarly, Kozlowski et al. (2000) proposed that if team members' tasks are highly interdependent, training should focus on intact teams, while if their tasks are similar and can be simply pooled, team members can be trained as individuals.

Several recent meta-analyses attest to the effectiveness of team training in improving the knowledge, skills, and attitudes of teams (Salas et al., 1999; Salas, Cooke, and Rosen, 2008; Klein et al., 2009; Delise, Gorman, and Brooks, 2010). Salas, Cooke, and Rosen (2008) examined the impact of specific team training on various outcome measures (i.e., affective, cognitive, process, and performance) and found that team training had a moderate, positive impact on team process ($\rho = .44$) and performance ($\rho = .39$).

These findings were further supported by another team-training meta-analysis that found that, in general, team training had positive effects (Delise, Gorman, and Brooks, 2010). This meta-analysis suggests that training may be more effective for learning when individuals have the opportunity to use the learned skills in the transfer environment. This is particularly promising for training of science teams and groups, suggesting that trainees could integrate the target skills into their daily activities to improve cognitive processes, such as deep knowledge integration, that leads to improved scientific performance (Salas and Lacerenza, 2013).

Team building is another intervention designed to improve overall team performance (Shuffler, DiazGranados, and Salas, 2011). Team building targets the interpersonal aspect of teamwork with particular emphasis on social interaction (Dyer, Dyer, and Dyer, 2007). Studies of team building have shown that it is not as effective as team training (Salas et al., 1999).

PROMISING PROFESSIONAL DEVELOPMENT INTERVENTIONS

Fiore and Bedwell (2011) elaborated the work of Cannon-Bowers et al. (1995) to propose a competency framework to support research on professional development (training) of science teams (see Table 5-1).

In science teams, context-driven competencies are those related to a particular research project. Such competencies can be developed through training focused on project goals, research tasks, and methods. Team-contingent competencies are those related to teamwork among these particular scientists and/or stakeholders and may be especially helpful to address

TABLE 5-1 Types of Team Competencies

Representative Science Team Competencies		Relation to Task	
		Task-Specific	Task-Generic
Relation to Team	Team-Specific	CONTEXT DRIVEN • Knowledge—Team objectives and resources • Skills—Particular analyses • Attitudes—Collective efficacy	TEAM CONTINGENT • Knowledge—Teammate characteristics • Skills—Providing teammate guidance • Attitudes—Team cohesion
	Team-Generic	TASK CONTINGENT • Knowledge—Procedures for task accomplishment • Skills—Problem analysis • Attitudes—Trust in technology	TRANSPORTABLE • Knowledge—Understanding group dynamics • Skills—Communication and assertiveness • Attitudes—Interdisciplinary appreciation

SOURCE: Adapted from Fiore and Bedwell (2011). Reprinted with permission.

challenges emerging from two features of team science—high diversity of team membership and high task interdependence. Team-contingent competencies can be developed through cross-training, in which individuals learn about the skills and duties of their teammates related to accomplishing scientific and/or translational tasks (see further discussion of cross-training below). For example, the Koch Institute for Integrative Cancer Research at Massachusetts Institute of Technology provides ongoing professional development opportunities to develop context-contingent knowledge of its particular research and translational mission and team-contingent competencies among its particular staff of life scientists, engineers, physicians, and other experts (see Box 5-1). Task-contingent competencies are those related to particular research tasks, such as experimental procedures. Finally, transportable competencies, useful across multiple science teams and/or larger groups, include such skills as mutual performance monitoring, giving and receiving feedback, leadership, management, coordination, communication, and decision making (Salas, Cooke, and Rosen, 2008).

This chapter now turns to a set of training strategies that show promise to address the coordination and communication challenges faced by science teams and larger groups. Many of these challenges can be addressed by developing team-contingent competencies, including "role knowledge"— understanding of the roles, tasks, skills, and knowledge each team member possesses. Coordination in science teams and groups can also be enhanced by developing context-driven competencies, including shared "mental models" (shared understandings of goals and tasks) among team members. Here, we discuss four research-based training strategies that show promise for enhancing coordination in science teams: cross-training, reflexivity training, knowledge development training, and team coordination training.

Cross-Training

Cross-training can help members of science teams or groups develop both knowledge of the roles and capabilities of diverse team members and also shared goals. Cross-training was developed to teach "interpositional knowledge" within a team, defined as a form of shared knowledge that includes understanding of task and role responsibilities of all team members, as well as understanding of the factors that influence the team and shared expectations about how the team will respond to changing environmental situations (e.g., Cannon-Bowers et al., 1998; Cooke, Kiekel, and Helm, 2001; Hollenbeck, DeRue, and Guzzo, 2004). Teams without such knowledge often suffer from coordination and communication problems (Volpe et al., 1996). Cross-training has been shown to improve the development of team interaction and shared mental models, which led to improved coor-

---

BOX 5-1
Professional Development for
Deep Knowledge Integration at the Koch Institute

The mission of the David H. Koch Institute for Integrative Cancer Research at Massachusetts Institute of Technology (MIT) can be briefly summarized as: "science + engineering = conquering cancer together" (see http://ki.mit.edu/ [April 2015]). This large group of scientists includes approximately 700 faculty, staff, and students within a 192,000-foot square building opened in the spring of 2011. Its research includes programs funded by the National Cancer Institute for multi-investigator grants in the areas of systems biology and cancer as well as nano-technology and cancer.

The institute's core intramural faculty consists primarily of biologists and engineers who formerly worked in different MIT departments, along with a small number of physician-scientists who both treat patients and have laboratories at the institute, students, and postdoctoral fellows in all of these fields. Through its "Bridge" project, the institute links its investigators to many more physician-scientists at area medical centers. The confluence of these multiple disciplines leads at times to "messy, turbulent waters" and a tower of Babel situation, according to Institute Director Tyler Jacks. However, the institute members are beginning to better understand each other, partly through participation in multiple, structured professional development opportunities. As shown in Figure 5-1, they include

- The Friday Focus seminar series, where graduate students and post-doctoral fellows join faculty mentors in presenting research methods and findings to the entire institute staff. For example, one seminar was humorously titled "Attack of the Layer-by-Layer Nanoparticles: Co-delivery of Chemodrug and RNAi for Cancer Treatment."
- Crossfire, a weekly educational series designed to bridge the biology/ engineering divide. The popular series was initiated by students and doctoral fellows, who both teach and attend the sessions in a peer-to-peer learning approach.
- A monthly lecture series, "The Doctor Is In," which helps scientists and engineers understand cancer through talks by physicians.
- An engineering "Genius Bar," created by postdoctoral fellows. Every 2 weeks, engineering fellows are available to answer questions on a speci-fied topic.
- An annual retreat for all staff with hundreds of presentations by institute members along with poster sessions.

From the perspective of the literature on team training (Fiore and Bedwell, 2011; see Table 5-1), these seminars, lectures, and discussions aim to develop context-driven competencies related to the institute's unique research and trans-lational mission and team-contingent competencies, including knowledge of other institute members' expertise and roles. The professional development opportuni-ties provide forms of cross-training that may help biologists and engineers to better understand and appreciate each other's skills, expertise, and duties related

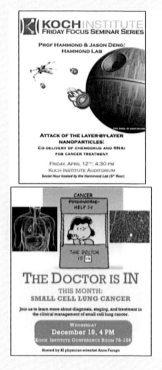

FIGURE 5-1 Posters illustrate some of the Koch Institute's professional development opportunities.
SOURCE: Presentation by Tyler Jacks to the committee, July 2013. Reprinted with permission.

to accomplishing shared research tasks and goals (see further discussion of cross-training below). As noted in Chapter 3, shared understanding of other team members' expertise and roles, referred to as "transactive memory" has been shown to enhance team effectiveness.

SOURCE: Presentation to the committee by Tyler Jacks, director of the Koch Institute. See http://www.tvworldwide.com/events/nas/130701/default.cfm, click on "Why Team Science" [April 2015].

dination and backup behaviors, and, consequently, improved performance (Marks et al., 2002) and team decision making (McCann et al., 2000).

Three types of cross-training methods are commonly used: (1) positional clarification, in which individuals are told about the other positions on their team; (2) positional modeling, in which individuals are both told about the position and have the opportunity to observe or shadow the position, thus gaining a deeper understanding of the duties involved; and (3) positional rotation, in which individuals are given hands-on training in the other positions such that they are able to perform the role if needed (Salas, Cooke, and Rosen, 2008; Klein et al., 2009; Delise, Gorman, and Brooks, 2010). Positional rotation was shown to improve teamwork knowledge and overall team performance over more traditional procedural or rule-based training in a simulated team environment (Gorman, Cooke, and Amazeen, 2010).

Positional rotation of investigators is generally not practical within an interdisciplinary or transdisciplinary science team or larger group, as learning to perform another's job would require obtaining an advanced degree in an unknown discipline. Nonetheless, more narrowly focused forms of cross-training, targeting the understanding of the roles, tasks, and expertise of team or group members, are feasible. Many of the courses and seminars offered at the Koch Institute are designed to help engineers and life scientists learn about others' roles, tasks, and expertise through direct engagement with each other. They go beyond positional clarification, in which an outside trainer or facilitator tells team members about others' roles, and are similar to positional modeling, in which the trainee observes or shadows a team member to learn about her or his role. For example, the engineering genius bar is an opportunity for life scientists, physicians, or other institute experts to directly observe engineers and ask questions about their work. Cross-training supports the development of not only shared mental models (Marks et al., 2002)—a team process known to enhance team performance—but also "transactive memory," or individuals' knowledge of the specializations of team members. Research on new and hybrid cross-training approaches could help address the question of how much knowledge of other disciplines is sufficient for proficient engagement in team science.

Team Reflexivity Training

Team reflexivity training, if adapted and translated to science contexts, is likely to help science teams and groups develop positive processes such as team self-regulation and team self-efficacy, facilitating the complex coordination of work required for success. In a review of methods for improving science collaboration, Salazar et al. (2012) suggested that enhancing reflexivity in science teams can improve team creativity as well as integration of individual member's knowledge. As discussed in Chapter 3, the life cycle of a team has been conceptualized in terms of episodes of planning, action, and reflection. Team reflexivity training requires members to reflect on prior performance, considering which objectives were or were not met, the strategies used or the group processes engaged, and how performance could be improved in the future, with the goal of improving future interaction (Gurtner et al., 2007). Reflections are prompted by a series of questions for team discussion, without the use of a facilitator or trainer, making this form of training relatively brief and inexpensive. Gurtner et al. (2007) found that teams receiving reflexivity training developed shared mental models to a greater extent than a control, with a positive impact on collaborative performance. In another study, van Ginkel, Tindale, and van Knippenberg (2009) found that reflexivity training improved shared team understanding of tasks and decision quality.

Similar to reflexivity training, in self-correction training, participants are empowered to improve their performance by reflecting on past performance episodes and self-diagnosing areas for improvement. Whereas reflexivity training is generally applicable to any setting and can be facilitated by a series of questions without the use of a facilitator or trainer, self-correction training requires more initial training for proper use. Because self-correction training is more focused and specific than reflexivity training, it has the potential for greater benefits (Gurtner et al., 2007). Guided team self-correction, or team dimensional training, is a specific type of self-correction that was derived from an expert model of teamwork, and has been found to improve both taskwork and teamwork performance (Smith-Jentsch et al., 2008). As noted in Chapter 1, this approach has been generalized and found to improve team members' shared mental models of teamwork across a variety of settings. It has been shown to increase performance and decrease errors in complex tasks such as naval submarine training simulations (Smith-Jentsch et al., 1998, 2008; Smith-Jentsch, Milanovich, and Merket, 2001).

Knowledge Development Training

Science teams and groups are composed of individuals with distinct sets of knowledge and expertise, which require integration to facilitate effective collaborative performance. This can be problematic given that research finds that different mental models of the task and the tendency to discuss commonly held information, as opposed to an individual's unique information, reduce performance. To address these problems, Rentsch et al. (2010) conducted a study explicitly focused on team training for knowledge building. Teams of undergraduates were trained to engage in communicative processes that elicit the structure and organization of their knowledge related to a team task designed by Navy Sea Air Land (SEAL) teams, as well as the assumptions, meaning, rationale, and interpretations associated with each member's knowledge.[2] The students used an external representation (i.e., an information board) that allowed team members to post, organize, and visually manipulate their knowledge related to the team task, more easily remember it, and draw attention to specific information as appropriate. The results showed that the knowledge-building training led to improved knowledge transfer (i.e., the exchange of knowledge from one team member to another), knowledge interoperability (i.e., shared knowledge that multiple team members are able to recall and use), cognitive congruence (i.e., an alignment or matching of team member cognitions), and higher overall team performance on the task (Rentsch et al., 2010).

In a follow-up study, Rentsch et al. (2014) tested a team-training strategy aimed at facilitating team knowledge-building in distributed teams. The authors found that teams trained to build knowledge, relative to untrained teams, shared more unique information, transferred more knowledge, developed higher cognitive congruence, and produced higher-quality solutions to a realistic problem-solving task.

Knowledge development training shows promise for improving collaborative problem-solving in science teams, by improving both knowledge building and knowledge sharing. However, other more general training strategies, such as reflexivity training and team development training, also improve knowledge building and knowledge sharing and, in addition, provide guidance in performance episodes.

Team Coordination Training

Team coordination training is a promising approach to facilitate the complex coordination of tasks required for success in science teams. This training was developed specifically to help teams modify responses based upon changes to their environmental situation. It focuses on helping teams adapt to the environmental demands of high-workload and time-stressed settings. This includes preplanning, information transmission, and anticipating information needs (Entin and Serfaty, 1999). It is primarily taught using vignettes to help teams recognize effective and ineffective teamwork. Practice and feedback are then provided in sessions where teams are able to apply what they have been taught and modify applications based upon errors. The goal is to turn explicit interaction factors that are thought to require effort on the part of the team (e.g., requests for information) into

---

[2] Open communication about assumptions and meanings underlying one's knowledge is also an element of the Toolbox intervention for interdisciplinary science teams and groups discussed later in this chapter.

implicit factors (e.g., providing information without being asked), in order to improve coordination. Although team coordination training was developed to help teams in contexts of high workload and stress, the competencies it develops (e.g., preplanning, anticipating information needs) are also suitable for teams in other contexts.

Gorman, Cooke, and Amazeen (2010) also explored a form of coordination training using methods described earlier in the cross-training section. The authors examined how to improve adaptability in teams through training that included disruptions to learned team coordination mechanisms. This involved, for instance, disrupting communication channels the team used to coordinate. Gorman, Cooke and Amazeen (2010) argued that this process-oriented training method helped teams deal with variability in coordination demands. Teams with disruption or "perturbation" training responded more adaptively to novel events than those with either cross-training or procedural training. The authors suggested that, similar to learning research on variability in practice, this helped teams generalize adaptive processes. By introducing coordination variability to the training, teams learned how to adapt their responses to changes in their environment and improve coordination during performance episodes. Science teams and larger groups face uncertainties that can arise from research findings (e.g., unanticipated results) or resource issues (e.g., loss of, or damage to, equipment; reduced grants) and hence might benefit from similar training approaches to increase their responsiveness to rapidly changing conditions.

NEW PROFESSIONAL DEVELOPMENT INTERVENTIONS FOR TEAM SCIENCE

Professional development designed specifically for science teams and groups is beginning to emerge, but only a few studies have examined its effectiveness for developing the targeted competencies or for improving performance. First, with support from the National Institutes of Health (NIH), the Northwestern University Center for Applied and Translational Sciences Institute developed an online training website, "TeamScience.net." The website includes a series of learning modules, message boards, and linked resources that aim to enhance skills for participating in or leading interdisciplinary and transdisciplinary science teams or groups. Two expert users (an academic medical doctor and a medical librarian) reviewed the website, finding that it followed principles of instructional design for adult education, was easy to navigate, and used attractive audiovisuals to present lessons, along with links to additional information and outside websites (Aranoff and Bartkowiak, 2012). But research to date has not included careful study of the website's learning goals and the outcomes for the users.

Second, the Toolbox Project (see http://toolbox-project.org [April 2015]), supported by the National Science Foundation (NSF), is a training intervention designed to facilitate cross-disciplinary communication in science teams and groups. O'Rourke and Crowley (2013) developed the Toolbox instrument to facilitate philosophical dialogue about science and the Toolbox workshop as a place for that dialogue. The instrument includes 34 probing statements accompanied by Likert scales to indicate the extent to which a respondent agrees with each statement. The statements are designed to elicit fundamental assumptions about science, including statements about ways of knowing (epistemologies), values, and the nature of the world. At the workshops, participants first complete the instrument and then engage in a facilitated dialogue lasting about 2 hours. At the end of the dialogue, they again complete the instrument. Data obtained from the workshop, including an audio file and pre- and post-dialogue reactions to the statements, are provided to the participants for analysis and reflection.

Although both the Toolbox instrument and the workshops are based on extensive theory and research and appear to target knowledge, skills, and attitudes supportive of interdisciplinary communication, to date there has been no empirical evaluation of whether participation in a Toolbox workshop leads to sustained improvement in cross-disciplinary dialogue after the workshop is over. In partial answer to this issue, Schnapp et al. (2012) analyzed data from a post-workshop survey that has been administered to 35 of the 90 teams and groups that have participated in a workshop. Just over half of those surveyed provided responses, and of these, 85 percent indicated that the workshop increased their awareness of the knowledge, opinions, or scientific approach of teammates, while 77 percent reported that the workshop had made a positive contribution to their professional development. A modified instrument for the health sciences was pilot-tested in two workshops with 15 participants, 10 of whom completed pre- and post-workshop questionnaires (Schnapp et al., 2012). Comparison of pre- and post-questionnaires revealed changes in about 30–40 percent of the items, related to motivations, research approaches, methods, confirmation, values, and reductionism, suggesting that the dialogue had met its goal of encouraging participants to thoughtfully consider other points of view.

EDUCATION FOR TEAM SCIENCE

Basic mastery of science concepts, methods, and perspectives provides the foundation for team science. In the 1960s and 1970s, when health sciences faculty experimented with interdisciplinary courses that focused on broad skills, curriculum committees and professional associations responded by mandating minimum levels of disciplinary knowledge and skills (Fiore, 2008). Reflecting such concerns, we first discuss science, technology, engineering, and mathematics (STEM) education in this section of the chapter, before turning to a discussion of interdisciplinary education.

STEM Education

Historically, science education has rarely prepared future scientists with the knowledge and skills required for effective knowledge integration and collaboration within a science team or larger group. Elementary and secondary school science classes typically ask students to work alone, listening to lectures, reading texts, or taking tests designed to measure recall of facts. There are few opportunities to learn to collaborate effectively or understand science as a social and intellectual process of shared knowledge creation (National Research Council, 2006, 2007b). At the undergraduate level, students majoring in science and the related STEM disciplines take courses dominated by lectures and short laboratory activities that often leave them with major misconceptions about important disciplinary concepts and relationships (National Research Council, 2006, 2012b).

At the doctoral level, some students participate in science teams and groups, but continue to receive little or no guidance or instruction on how to be an effective collaborator. Students develop deep conceptual understanding of topics and methods within a discipline, and are trained in its unique perspectives, languages, and standards of evidence (epistemologies). As a result, they may consciously or unconsciously develop a negative perception of other disciplines (National Academy of Sciences, National Academy of Engineering, and Institute of Medicine, 2005). The hallmark of doctoral education is the student's individual, unique, and original research, and teamwork at this stage may be actively discouraged (Nash, 2008; Stokols, 2014).

Collaborative Education in STEM Classrooms

New developments in K-12 and higher education potentially could enhance preparation for team science, developing both disciplinary and interdisciplinary knowledge and collaborative skills. The NRC *Framework for K-12 Science Education* (2012c) draws on decades of research showing that engaging students in science practices—such as asking questions, developing and using models, or engaging in argument from evidence—helps them master science concepts and facts (National Research Council, 2007b). Although students often work in small groups when engaging in these science practices, instruction has not been designed to integrate development of collaboration skills along with STEM concepts and skills.

Collaborative learning activities are also being tested in higher education. Research has shown that undergraduate learning of STEM is strengthened when students work collaboratively to solve problems, reflect on laboratory investigations, and discuss concepts and questions (National Research Council, 2012b). However, these approaches have not been widely adopted by college faculty, and, as at the K-12 level, they focus primarily on acquisition of STEM content and skills, with little attention to collaborative skills.

Gabelica and Fiore (2013a) reviewed studies of three group learning interventions in STEM higher education: problem-based learning, team-based learning, and studio learning. In all three approaches, faculty members present students with an authentic problem or assignment and students work in small groups to understand the issues at hand, gather relevant information, and develop solutions. All three approaches have been shown to enhance students' understanding of targeted STEM concepts and skills under certain conditions (Gijbels et al., 2005; Strobel and van Barneveld, 2009), and a few studies of team-based learning also reported gains in students' interpersonal and teamwork skills (e.g., Hunt et al., 2003). However, interpersonal and teamwork skills were seldom measured, partly because students were sometimes reluctant to rate their peers' contributions to the team's work (Thompson et al., 2007).

Gabelica and Fiore (2013b) recommended ways to address this gap, suggesting that developers of such interventions integrate insights from the organizational research on teams. This would involve, for example, assessing students' development of interpersonal teamwork skills through self-ratings of interpersonal skills (Kantrowitz, 2005) and behaviorally oriented rating scales for self- and peer-evaluations of contributions to the team (Ohland et al., 2012).

Borrego et al. (2013) also recommended that developers of group learning interventions draw on the team's research. In a two-phase study, the authors first reviewed 104 articles describing student team projects in engineering and computer science. They found that faculty assigned team projects to advance diverse learning goals, including teamwork, communication skills, lifelong learning, sustainability, and professional ethics. The student teams experienced team process challenges (e.g., conflict), and faculty members tried to address these challenges as they arose but were not aware of methods from the organizational literature that could be used to illuminate the very challenges they had sought to address. Second, Borrego et al. (2013) reviewed the organizational literature related to five team processes identified as important in the studies of student teams, clarified how these processes impacted student success, and developed theories of team effectiveness specific to engineering education.

Finally, research by Stevens and Campion (1994) has identified transportable individual competencies required for effective teamwork, showing promise for use within collaborative STEM education. These authors not only explicated teamwork competencies but also developed and validated the Teamwork Test (Stevens and Campion, 1999) for measuring these competencies.

In sum, research to date has shown that carefully designed educational interventions that engage students in small group investigations, discussion, and problem-solving activities can support STEM learning, but has not yet examined the potential of such small groups to also serve as contexts for learning teamwork skills. Integration of concepts and methods from the organizational sciences with STEM education could redress this gap.

Interdisciplinary and Transdisciplinary Higher Education

Stokols (2014) observed that science teams and groups often address the coordination and communication challenges arising in interdisciplinary or transdisciplinary research by drawing on online resources and/or providing training on an ad hoc basis. He proposed that longer-term education is needed to prepare a generation of scholars capable of addressing complex scientific and societal challenges in collaborative, interdisciplinary or transdisciplinary research environments. Consideration of this proposal is informed by reflecting on the United States' long history of interdisciplinary education, as well as more recent courses and programs focusing specifically on team science.

As the health sciences began to develop interdisciplinary programs in the 1960s (Lavin et al., 2001), researchers were prompted to address the communication and teamwork challenges inherent in these educational approaches (Hohle, McInnis, and Gates, 1969). This led to the creation of interdisciplinary internships and fellowships designed to help students learn to communicate across disciplines (Lupella, 1972) and highlighted the need for research and training related to the development of collaboration skills in team settings (Jacobson, 1974). Although interdisciplinary education grew over the following decades, knowledge of how to support development of collaboration and teamwork skills remained relatively static (Fiore, 2008).

Interdisciplinary education has grown rapidly over the past four decades (Lattuca et al., 2013). Between 1975 and 2000, the number of interdisciplinary majors at U.S. colleges and universities increased by 250 percent, a period when student enrollments increased only 18 percent. However, colleges and universities have been slow to support this shift toward interdisciplinary teaching and learning with supportive formal policies and practices. Klein (1996) called on universities to support faculty professional development in interdisciplinary teaching and to protect faculty from discipline-centric norms, such as tenure reviews that punish work outside one's discipline. She suggested that such supports as mentoring, physical space for collaborations, and cross-disciplinary training would help to develop new norms of interdisciplinarity. More recently, Klein (2010) argued that, to sustain interdisciplinary teaching and learning, institutional support must be consistent and embedded within the university culture.

Defining Competencies for Team Science

A critical issue is the lack of conceptual clarity about the learning goals of interdisciplinary and transdisciplinary education that aims to prepare students for team science. Researchers have proposed a variety of team science competencies as important learning goals for such education. We next discuss these competencies and provide a clustering of them in Table 5-2. More problematic is the lack of prospective experimental or quasi-experimental studies of learning outcomes, as the research has relied heavily on surveys, interviews, and archival analyses.

Building on an earlier framework by Stokols et al. (2003), Nash and colleagues (2003) delineated three types of core competencies for the transdisciplinary scientist: (1) attitudinal; (2) knowledge; and (3) skill-based. They proposed that all three types could be developed through graduate and postgraduate education, including coursework, seminars, and workshops taught by disciplinary and interdisciplinary faculty; mentoring by research supervisors in multiple disciplines; group work with other transdisciplinary trainees, such as a journal club; and a supportive institutional environment.

Using a consensus study of expert opinion, Holt (2013) identified a somewhat similar list of competencies for effective performance in team science contexts and recommended that they be developed in graduate education through interdisciplinary coursework and seminars along with team research and projects. Borrego and Newsander (2010) developed another list of competencies in a study of the NSF Integrative Graduate Education and Research Traineeship (IGERT) Program, which supports training of scientists for interdisciplinary team science. The authors grouped the diverse learning outcomes articulated across 130 successfully funded proposals, as follows:

- Disciplinary grounding: Although the awards are interdisciplinary by definition, a full 50 percent of proposals argued that graduate student trainees would gain grounding in a specific discipline.
- Teamwork: The most clearly articulated learning outcome, in 41 percent of the proposals, was that the proposed center would create a culture of teamwork.

TABLE 5-2 Competencies for Productive Participation in Team Science

	Competency	Examples	References
Values, Attitudes, and Belief-Based Competencies	Valuing Interdisciplinary or Transdisciplinary Collaboration	Attitudes that predispose one to integrate knowledge from a varied set of disciplines The beliefs that such efforts are necessary and can lead to effective outcomes	Nash et al. (2003); Klein, DeRouin, and Salas (2006); Nash (2008); Fiore (2013); Stokols (2014); Vogel et al. (2014)
	Societal and Global Perspectives	Belief that complex problems should be approached from a broad, multilevel perspective	Borrego and Newsander (2010); Stokols (2014)
	Collaborative Orientation	Values that emphasize inclusion of multiple and diverse perspectives	Klein, DeRouin, and Salas (2006); Hall et al. (2008); Fiore (2013); Stokols (2014)
Knowledge-Based Competencies	Understanding Other Disciplines	Understanding core theories and methods from other disciplines	Nash et al. (2003); Nash (2008)
	Disciplinary Awareness and Exchange	Awareness of assumptions of own discipline, engage colleagues from outside disciplines Skills and knowledge to think across disciplines and synthesize varied concepts and theories	Schnapp et al. (2012); Holt (2013); Lattuca et al. (2013); Stokols (2014)
	Processes of Integration, Integrative Capacity	Develop shared interdisciplinary vision, modify work based upon influence of others	Marks et al. (2002); Borrego and Newsander (2010); Salazar et al. (2012); Holt (2013)
	Disciplinary Grounding	Cultivation of deep knowledge within one or more disciplines	Borrego and Newsander (2010)

Continued

TABLE 5-2 Continued

	Competency	Examples	References
Interpersonal/Skill-Based Competencies	Scientific Skills Across Disciplines	Use theories and methods of multiple disciplines	Gebbie et al. (2007); Vogel et al. (2012)
	Methodology	Taking a methodologically pluralistic approach	Nash et al. (2003); Nash (2008)
	Teamwork and Taskwork	Knowledge of resources and strategies to enhance teamwork as well as taskwork	McCann et al. (2000); Salas, Cooke, and Rosen (2008); Smith-Jentsch et al. (2008); van Ginkel, Tindale, and Van Knippenberg (2009); Borrego and Newsander (2010); Gorman et al. (2010); Holt (2013)
	Interdisciplinary Research Management	Develop team skills to strengthen team structure and dynamics	Holt (2013)
	Leadership	Build communication strengths, manage conflict, trust the value of teammates	Bennett and Gadlin (2012); Holt (2013); Ekmekci, Lotrecchiano, and Corcoran (2014)
	Fruition	Presenting research at interdisciplinary conferences, partner with those in other disciplines on proposals	Holt (2013)
	Interdisciplinary Communication	Active listening, oral and written, assertive communication / Communicate regularly with scholars from other disciplines	Klein, DeRouin, and Salas (2006); Gebbie et al. (2007); Borrego and Newsander (2010); Fiore (2013)
	Interact with Others	Engage colleagues from other disciplines	Gebbie et al. (2007); Vogel et al. (2014)
	Coordination	Capacity to adapt flexibly and effectively to situational and intra-team challenges	Entin and Serfaty (1999); Klein, DeRouin, and Salas (2006); Gorman et al. (2010); Fiore (2013)
	Interdisciplinary Skills	Ability to consider and apply perspectives from outside one's discipline	Lattuca et al. (2013)

TABLE 5-2 Continued

	Competency	Examples	References
	Transdisciplinary Behaviors	The behaviors that support activities for integrating perspectives and working with others outside one's discipline	Stokols (2014)
	Intellect and Self-Awareness	Broad intellectual curiosity, recognition of personal strengths and weaknesses with regard to interdisciplinary research	Hall et al. (2008); Holt (2013)
	Reflective Behavior	Ability to recognize when one's general approach, or a specific problem-solving approach, needs to be changed	Lattuca et al. (2013); Stokols (2014)
	Critical Thinking	Critical awareness about one's own potential disciplinary biases in collaborative situations	Borrego and Newsander (2010); Hall et al. (2012a); Vogel et al. (2014)

Intrapersonal-Based Competencies (vertical label)

- Integration: Thirty percent of the proposals argued that their graduate programs would encourage students to integrate concepts from relevant disciplines.
- Societal and global perspectives: Twenty-four percent of the proposals noted that they would encourage students to consider societal and global issues.
- Interdisciplinary communication: Twenty-four percent of the proposals noted that their projects would emphasize the importance of interdisciplinary communication.

Borrego and Newsander (2010) also found that scientists, engineers, and scholars in the humanities had different views of "integration." For scientists and engineers, "teamwork" was fundamental, whereas scholars in the humanities considered "critical thinking" as more central. The authors suggested that because critical reflection on disciplinary inconsistencies and limitations is a particular strength when solving complex interdisciplinary problems, scientists and engineers should incorporate critical thinking as a goal of interdisciplinary education.

Engineering students are often assigned to work in interdisciplinary teams, and Lattuca, Knight, and Bergom (2013) developed a self-report measure of interdisciplinary engineering competence, including three scales: interdisciplinary skills, reflective behavior, and recognizing disciplinary perspectives. Importantly, the scales do not include any measures of teamwork or interpersonal skills. Lattuca, Knight, and Bergom (2013) caution that the scales are preliminary and that they were unable to evaluate their construct validity (their relationship to the target competencies), "because direct measures of interdisciplinary knowledge and skills do not exist" (p. 737).

Gebbie et al. (2007) identified competencies for transdisciplinary health research. Using a Delphi technique to elicit information from several groups of experts in interdisciplinary research and education, they arrived at 17 statements describing what a well-trained scholar should be able to do when participating in interdisciplinary research. The statements were grouped into three categories: conduct research, communicate, and interact with others.

As discussed above, Cannon-Bowers et al. (1995) suggested "transportable team competencies" as a focus for educational programs to develop the kinds of competencies that can be applied across different tasks and teams. Building on this, as well as a framework of interpersonal skills created by Klein, DeRouin, and Salas (2006), Fiore (2013) developed a framework of transportable interpersonal competencies for team science. This framework specified the forms of active listening, oral and written communication, assertive communication, relationship management competencies, coordination, interdisciplinary appreciation, and collaborative orientation that support effective collaboration in science. Fiore suggested that these competencies be integrated as learning goals for interdisciplinary education to support team science.

Stokols (2014) conceptualized a broad intellectual orientation for transdisciplinary team science including values, attitudes, beliefs, skills and knowledge, and behaviors (see Table 5-2). Both Stokols (2014) and Misra et al. (2011a) emphasized the role of mentors in graduate education, noting that mentors who encourage the acquisition and synthesis of a broad knowledge base can help students acquire the skills and attitudes foundational to transdisciplinary work. Stokols (2014) also suggested that, when students are trained in institutions that engage them in authentic team science research activities focused on real-world problems, "they are better able to avoid the conceptual biases associated with *disciplinary chauvinism* and the *ethnocentrism* of traditional academic departments" (p. 66).

In sum, many authors have proposed various competencies for team science and educational strategies to develop these competencies, and there are areas of overlap and agreement within this variety. However, the research to date has not identified a common set of agreed-on competencies that could serve as targets for design of educational or professional development courses.

Research on Educational Interventions for Team Science

There have been only a few empirical analyses of educational strategies aimed at preparing individuals for team science. These educational strategies vary, including programs implemented within individual schools or universities as well as larger, federally funded education programs. In addition, the research to date has not examined how acquisition of the targeted competencies may enhance the effectiveness of science teams.

Graduate Education for Team Science

The University of California, Irvine's School of Social Ecology offers a doctoral seminar specifically developed to expose students to a broad range of relevant disciplines. To examine the influence of the seminar, Mitrany and Stokols (2005) conducted a content analysis of doctoral dissertations produced by the school, reporting that the dissertations provided evidence of an interdisciplinary orientation reflected, for example, in the multidisciplinary composition of their faculty committees and the cross-disciplinary scope of their research topics, conceptual frameworks, and multimethod analyses.

Carney and Neishi (2010) conducted an evaluation of the IGERT Program described above, using surveys and data from IGERT graduates and a control group of doctoral graduates from similar academic departments that did not participate in the program. In comparison to the non-IGERT graduates, a higher percentage of IGERT graduates reported that they drew on at least two disciplines in their dissertation research and obtained their doctoral degrees in less time (thanks to the program's financial support). Contrary to some previous authors who warn that interdisciplinary doctoral students may face challenges in the discipline-based academic job market (e.g., Nash, 2008), the IGERT graduates reported that their interdisciplinary research training and the program's professional networking opportunities gave them a competitive edge in the job market. They reported less difficulty acquiring their first jobs than the non-IGERT graduates. At these jobs, they were significantly more likely than their non-IGERT peers to conduct research or teach courses that require integration of two or more disciplines.

In a separate study of the IGERT Program, Borrego and colleagues (2014) sought to identify longer-term outcomes of the traineeships for the host universities as well as the trainees, by interviewing faculty and administrators at a small number of institutions. The interviewees reported overcoming barriers to successful implementation of the interdisciplinary doctoral training program through, for example, changes to eligibility criteria for advisers so that faculty from varied departments could serve as a doctoral student's adviser. In addition, departments changed their policies to reward faculty for advising outside their department, and some institutions expanded eligibility for fellowships so that students from interdisciplinary programs could compete for the awards. In addition, many programs created interdisciplinary courses or seminars and required that students participate in team research and take laboratory classes from different disciplines.

The National Cancer Institute's Transdisciplinary Research on Energetics and Cancer I (TREC I) project sought to develop three types of competencies for graduate students (Vogel et al., 2012):

1. scientific skills, including educational grounding in two or more disciplinary perspectives and skills for integrating and synthesizing approaches across disciplines;
2. intrapersonal skills, including positive attitudes, values, and beliefs about the transdisciplinary approach and critical awareness of the relative strengths and limitations of all disciplines (referred to as a transdisciplinary orientation); and
3. interpersonal skills for collaborating and communicating across disciplines, such as the ability to use analogies, metaphors, and lay language in lieu of discipline-specific jargon and willingness to engage in continual learning.

The four TREC centers implemented a variety of training activities to develop these competencies, including interdisciplinary research courses, journal clubs, and writing retreats to develop skills in collaborative writing and research. Many centers also provided co-mentoring and multi-mentoring to expose trainees to multiple disciplinary perspectives, and a cross-center working group developed additional training activities.

Multiple mentors were expected to play a key role in developing the three types of competencies, by teaching trainees about the concepts, theories, and methods of the different disciplines; facilitating learning of interpersonal skills for transdisciplinary research; and providing support for career advancement (e.g., the mentors would provide visibility to and coach the trainee). The "multi-mentoring" approach was also expected to provide social support and role modeling. However, each TREC center was allowed to develop its own training program, and the study found that only about 60 percent of trainees had two or more mentors.

An analysis of these training activities found gains in all three types of competencies, including students' attitudes toward working across disciplines, ability to work across disciplines, and scientific competency. Importantly, the trainees also improved in scholarly productivity, as measured by number of publications/presentations and number of collaborative authors. Multimentoring experiences were associated with greater transdisciplinary orientation and positive perception of one's center (Vogel et al., 2012).

Undergraduate Education for Team Science

Few studies have examined the goals and outcomes of interdisciplinary undergraduate programs focusing on team science. One example was a study of the University of California, Irvine's Interdisciplinary Summer Undergraduate Research Experience Program, which aims to develop students' ability to integrate research concepts and methods. Misra et al. (2009) examined curriculum strategies (such as the use of team projects, research, or journal club meetings), interdisciplinary processes (such as student participation in team projects), and student products (completed projects, papers, and course grades) for a group of participants. Over the course of the program, participants developed more positive attitudes toward interdisciplinary research and participated in interdisciplinary research activities more frequently. In comparison with another group of students who participated in a different research fellowship program that did not include an interdisciplinary component, the participants showed no significant difference in student products, but a higher level of engagement in interdisciplinary collaborative research. Further, team-focused projects were found to be instrumental to these changes.

In light of Borrego and Newsander's (2010) suggestion that critical thinking is valuable for interdisciplinary collaboration in science and engineering, a recent study by Lattuca et al. (2013) focused on this competency. In a longitudinal study of about 200 students, the authors compared students majoring in traditional disciplinary programs with those participating in interdisciplinary programs, using existing assessments of critical thinking, need for cognition, and attitudes towards learning. They found no significant differences in levels of these competencies between the two groups that could be attributed to major or structure of the program.

The Role of Mentoring for Team Science

The research discussed above consistently identifies mentoring as a crucial component of interdisciplinary education for team science, but only a few programs focus specifically on mentoring. For example, NIH's Building Interdisciplinary Research Careers in Women's Health Program is designed for junior faculty interested in advancing research in women's health. The program establishes mentoring teams to provide the young faculty members with multiple perspectives on a range of scientific and career issues. A recent study showed that a majority of scholars in the program had applied for competitive grants after completing the training and that approximately half were successful (Nagel et al., 2013).

In 2010, NSF adopted a new policy requiring that requests for funding of postdoctoral researchers include a postdoctoral researcher mentoring plan. Implemented in part to advance NSF's two core strategies of fostering the integration of education and research and expanding the participation of groups and institutions that have been underrepresented in science, the plans must describe mentoring activities, such as career counseling, training in preparation of grant proposals and publications, and "guidance on how to effectively collaborate with researchers from diverse backgrounds and disciplinary areas" (National Science Foundation, 2014b). Recent reports, although anecdotal, suggest that reviewers of NSF proposals may be placing increased weight on this requirement (Flaherty, 2014).

Currently, however, mentoring, and especially interdisciplinary mentorship, is too often lacking for students and scholars. In a recent survey on the "Global State of Young Scientists," the unavailability of mentoring was one of the top four career obstacles identified (Friesenhahn and Beaudry, 2014). Survey responses indicated that junior scientists are not explicitly taught how to train and supervise students and postdoctoral fellows, but are expected to learn by experience.

ADDRESSING THE SEVEN FEATURES THAT CREATE CHALLENGES FOR TEAM SCIENCE

In this section, we consider how the research reviewed in this chapter may help guide professional development, training, or education for team science as a way to address the communication and coordination challenges that emerge from the key features that create challenges for team science.

High Diversity of Membership

The challenges of communication and interpersonal interactions raised by high diversity of team membership can be addressed in part with cross-training and other types of training focusing on team-specific competencies, to help team members better understand and appreciate the varied knowledge and roles of different team members. These challenges also can be addressed through interdisciplinary educational seminars that expose team members to scholars from different disciplines, such as those offered by the Koch Institute or through structured approaches such as the Toolbox workshops described above. In addition, professional development or education for team science could focus directly on development of interpersonal skills such as "active listening" with the goal of ensuring that inputs from those in different disciplines are understood.

Deep Knowledge Integration

As noted in Chapter 1, science teams and groups that seek to deeply integrate, or even transcend, the knowledge of individuals who may have different goals, assumptions, and languages often encounter communication and coordination challenges. Professional development focused on developing shared understanding of each member's knowledge—such as cross-training, knowledge sharing training, and coordination training—may help to address these challenges. Education or professional development devised to illustrate larger connections across disciplines (both conceptual and methodological) also would help foster knowledge integration.

Large Size

Although training to develop shared knowledge of fellow team or group members' knowledge and skills can help to overcome the communication and coordination challenges raised by large size, this training may have to be relatively shallow. For example, cross-training may focus on positional clarification (knowledge of other members' roles), rather than deeper understanding of other members' knowledge, skills, and tasks, both because of the large number of members and because it is not practically possible to quickly develop deep understanding of an unfamiliar discipline. As a first step, leaders of large groups may consider engaging training experts to identify the amount of "interpositional knowledge" necessary to support behavioral coordination across the team.

Goal Misalignment with Other Teams

Lack of goal alignment with other teams may result partly from team members' lack of awareness of common goals and partly from organizational factors that are beyond the scope of team training. Training or professional development can be designed to increase awareness of the common goal and how the goals of the varied subteams are linked to that goal. In addition, this challenge can be addressed through reflexivity training. Teams that reflect on prior performance episodes can develop knowledge of when goal alignment and/or misalignment with other teams is affecting their interactions and performance. Educational interventions that include group activities, such as problem-based learning and team-based learning, also could introduce the concept of goal alignment to help students learn how to manage goal conflicts that often arise between different science teams.

Permeable Team and Group Boundaries

Permeable boundaries create a need for the context-driven, team-contingent, and task-contingent competencies shown in Table 5-1. In terms of the context, team or group members who are new to a project would need training in the project's scientific and/or translational goals. From the task standpoint, new members may require training in particular research methods or analyses to accomplish research tasks. From the team standpoint, transitional membership creates a gap in team-specific knowledge, as a new member may not understand teammates' expertise and roles. Such a gap could be addressed by cross-training or knowledge development training.

Geographic Dispersion

Geographic dispersion of team members necessitates training to develop team or group members' understanding of each other's expertise, roles, and context-driven and team-contingent competencies. Cross-training or knowledge development training may help to provide this understanding, thus facilitating coordination. However, because dispersion hinders acquisition of this understanding, training focused on development of team cohesion or team self-efficacy might also be beneficial. Reflexivity training can also be used to identify when and where proximity is creating problems for the team.

High Task Interdependence

The high level of interdependency within science teams and groups creates a need for both context-specific and team-specific knowledge. Because one or more members is likely to have the expertise needed to accomplish each piece of the research project (e.g., expertise in statistics), knowledge of different team or group members' expertise can facilitate coordination, supporting team effectiveness (Kozlowski and Ilgen, 2006). To develop context-specific competencies, training should focus on task-specific knowledge and skills. To develop team-specific knowledge, reflexivity training is a promising method. Both training strategies can support the deep integration of team members' knowledge needed to achieve the scientific and/or translational goals of the project.

SUMMARY, CONCLUSIONS, AND RECOMMENDATION

Research on teams in a variety of contexts outside of science has been applied to develop training strategies, shown to improve team processes and effectiveness. Several research-based training strategies show promise to enhance communication, coordination, and knowledge integration in science teams, overcoming the challenges that emerge from diverse membership, large sizes, high task interdependence, and other features of team science. The committee expects that translation and application of these strategies to create professional development programs for science teams would enhance the effectiveness of these teams. Professional development programs for team science are beginning to emerge, and these programs would benefit from translation and application of the strategies shown to enhance effectiveness in non-science contexts.

CONCLUSION. *Research in contexts outside of science has demonstrated that several types of team professional development interventions (e.g., knowledge development training to increase sharing of individual knowledge and improve problem solving) improve team processes and outcomes.*

RECOMMENDATION 2: **Team-training researchers, universities, and science team leaders should partner to translate, extend, and evaluate the promising training strategies, shown to improve the effectiveness of teams in other contexts, to create professional development opportunities for science teams.**

The TeamSTEPPS Program illustrates the approach the committee recommends to improve the training and performance of science teams. TeamSTEPPS extends and translates research findings on team effectiveness in aviation to create health care team-training practices with the goal of improving health care performance. The program was developed in response to the Institute of Medicine (1999) report *To Err Is Human: Building a Safer Healthcare System*, which identified the need to improve team performance in medical settings as one of several steps recommended to reduce medical errors and improve health care. As described by Alonso and colleagues (2006), the program's developers reviewed more than 20 years of research on teams and team performance to identify critical competencies needed for effective teamwork and translate them for health care contexts. The list of competencies was then converted into a framework of trainable team skills, and training strategies were developed to strengthen these skills.

Although research has demonstrated that training for current team members can increase team effectiveness, educational programs designed to prepare students for future team science have only recently emerged and have not yet been systematically evaluated. Further work is needed to more clearly specify the competencies needed for team science and to develop assessments of these competencies; such research would clarify learning goals, an important step toward enhancing learning outcomes. To date, there has been little empirical evaluation of which educational activities are most effective for developing particular competencies, nor whether, and to what extent, acquisition of these competencies contributes to the effectiveness of science teams or larger groups.

CONCLUSION. *Colleges and universities are developing cross-disciplinary programs designed to prepare students for team science, but little empirical research is available on the extent to which participants in such programs develop the competencies they target. Research to date has not shown whether the acquisition of the targeted competencies contributes to team science effectiveness.*

6

Team Science Leadership

This chapter begins with a discussion of the definition of leadership and the degree to which it is distinct from management. We then review the expansive parallel literatures on team and organizational leadership in contexts outside of science. Through the lens of established leadership theories, models, and behaviors, we identify those approaches that are relevant to science teams and larger groups and for which there is research evidence for enhanced team or group effectiveness. Next, we summarize the research evidence on team science leadership. We then discuss professional leadership development for team science leaders. We then use the research evidence as a guide to consider how leadership strategies can address the challenges for team science created by the seven features outlined in Chapter 1 and conclude with a conclusion and a recommendation for the future leadership of science teams and groups.

DEFINING LEADERSHIP AND MANAGEMENT

Our study charge calls for consideration of how different management approaches and leadership styles influence the effectiveness of team science. The distinction between management and leadership has been defined in the research literature in multiple ways. For example, Kotter (2001, p. 85) proposed that leadership and management are "two distinctive and complementary systems of action." Kotter (2001) proposed that the main functions of leadership are to set direction, to align people, and to motivate and inspire them, while the main functions of management are to develop concrete plans for carrying out work, to allocate resources appropriately, to create an organizational structure and staffing plan, and to monitor results and to develop problem-solving strategies when needed. However, Drath et al. (2008, p. 647) pointed out that these functions are not necessarily mutually exclusive: "alignment is often achieved through structure and many of the aspects of shared work usually categorized as management, such as planning, budgeting, supervisory controls, performance management, and reward systems." Recognizing that it is difficult, if not impossible, to draw a strict line between leadership and management, we have not attempted to completely disentangle the two functions. Therefore, while this chapter focuses primarily on leadership, the research discussed also addresses aspects of management (as defined by some scholars). Management of organizations that house science teams is discussed further in Chapter 8.

RESEARCH FINDINGS ON GENERAL LEADERSHIP AND POTENTIAL IMPLICATIONS FOR TEAM LEADERSHIP

Over half a century of research on leadership has highlighted the nuances and complexities of leading individuals, teams, and organizations. Some leaders are born with the skills and abilities to guide followers, while other leaders are trained through education and opportunities for hands-on experience. Those who lead large organizations successfully are not necessarily successful at leading small groups. Some leaders are charismatic and have a commanding presence in a crowd while other leaders build trust and respect through one-on-one relationships. In short, leadership is not a quality that an individual either has or lacks, and there is not a single leadership style that is effective in all contexts. Rather, leadership is multifaceted, encompassing different ways in which individuals exhibit leadership as well as different environments in which leadership occurs. Leaders' approaches to their team or group members may vary depending upon the nature of the task and goals for the team, as well as the composition of the team. In some cases, a directive, task-oriented approach may be called for, while in other cases, leaders strive to support and encourage team members' ideas, innovations, problem identification, and proposed solutions.

This chapter will show that researchers have focused on many aspects of leadership, including specific leader behaviors, their interactions with followers, and contingent factors that guide how effective a leader is in a given situation.

This general leadership theory and research can inform the emerging field of team leadership, yet it must be noted that leadership quality is very difficult to measure or evaluate; in the research to date, the most common criterion for leadership effectiveness is the subordinates' perception of the effectiveness of their leader, rather than direct measures of team performance. Nonetheless, meta-analytic findings from this extensive literature provide indications of the potential value of leadership in promoting team effectiveness (Kozlowski and Ilgen, 2006). In this section, we review the research evidence for the impact of *behavioral, relational, transformational, transactional, contingency,* and *contextual* approaches to leadership, with particular emphasis on contextual approaches. Each of these approaches entails different behaviors on the part of leaders (and in one case—the relational approach—also emphasizes the behavior of followers), but they are not necessarily mutually exclusive and a single leader can employ multiple approaches.

Behavioral Approach

Influential studies conducted at the Ohio State University in the 1950s identified two overarching features of a *behavioral* approach to leadership: consideration (i.e., supportive, person-oriented leadership) and initiating structure (i.e., directive, task-oriented leadership) (Day and Zaccaro, 2007). Team outcomes have been found to be significantly correlated with both features, suggesting that this classic approach is potentially viable for team leadership as well (Judge, Piccolo, and Ilies, 2004). An advantage of this behavioral approach is its focus on observable leader behaviors rather than personality traits, allowing many of its core elements of this approach to be used with other leadership approaches, especially the transformational approach, discussed below (Bass and Riggio, 2006).

Relational Approach

The *relational* approach, or leader-member exchange theory (LMX), describes the dyadic relationship between leaders and followers, or subordinates. Research shows that followers who negotiate high-quality exchanges with their leaders experience more positive work environments and more effective work outcomes (Gerstner and Day, 1997; Erdogan and Bauer, 2010; Wu, Tsui, and Kinicki, 2010). In this view, team leaders become especially important for shaping team members' perceptions of their shared environment and of team relationships (Kozlowski and Doherty, 1989; Hofmann, Morgeson, and Gerras, 2003).

Transformational Approach

The *transformational* approach, the most dominant leadership paradigm over the past decade, focuses on leadership styles or behaviors that induce followers to transcend their interests for a greater good (Kozlowski and Ilgen, 2006; Day and Antonakis, 2012). Transformational leadership encompasses the behavioral dimensions of charisma, inspirational motivation, intellectual stimulation, and individualized consideration.

While the transformational approach may be of particular relevance to teams, it has been studied mainly at the individual level of analysis, assessing how leaders using this approach influence individual followers[1] and outcomes rather than team-level outcomes. In one of the few studies looking specifically at teams, Lim and Ployhart (2004) found the transformational approach to be more strongly related to performance in maximal-performance than in typical-performance contexts, supporting the notion that transformational leadership facilitates subordinate motivation and effort.[2] Other studies have linked the transformational approach to facets of a team's collective personality and to its performance/profitability (Hofmann and Jones, 2005). Of direct relevance to science teams, recent research has demonstrated the multilevel and cross-level influences of transformational leadership on the effectiveness of innovation teams (Chen et al., 2013). In another example of the multilevel influences of organizational and team leadership, Schaubroeck et al. (2012) found that higher-level leaders influence lower-level leaders and teams by serving as leader models to emulate and by crafting cultures that influence the lower level via alternative pathways.

Transactional Approach

The *transactional* approach (Bass, 1985) entails leader behaviors aimed at negotiating mutually beneficial exchanges with subordinates. These behaviors can encompass contingent rewards, including clear expectations and linkages with outcomes, active management by exception (i.e., proactive and corrective action), and passive management by exception (i.e., reactive management after the fact).

Contingency and Contextual Approaches

The *contingency* approach matches the leader's behavior to the context to maximize outcomes and leadership effectiveness. This emphasis on context should be relevant to teams engaged in complex tasks, as is the case for science teams (Dust and Zeigert, 2012; Hoch and Duleborhn, 2013). While

---

[1] For leaders to exercise influence, followers must allow themselves to be influenced (Uhl-Bien and Pillai, 2007). For a discussion of followership theory and a review of research related to followership, see Uhl-Bien et al. (2014).

[2] Maximal-performance contexts involve tasks of relatively short duration in which team members are aware that performance is being evaluated and accept that that maximal performance is expected on the task (Sackett, Zedeck, and Fogli, 1988, as cited in Lim and Ployhart, 2004).

the contingency approach is no longer active in current research, it has been tied to the development of a *contextual* approach to leadership. As its name suggests, this approach emphasizes a more contextual perspective that recognizes the need to use a combination of approaches to meet the leadership requirements of particular situations (Hannah et al., 2009; Simonton, 2013; Hannah and Parry, in press). For example, the contextual circumstances of a particular team might require shared leadership, in which leaders share leadership roles, functions, and behaviors among team members. Shared leadership can be formally appointed at the outset of an endeavor or can emerge during the course of an activity (Mann, 1959; Judge et al., 2002). Leadership emergence involves both the extent to which an individual is viewed as a leader by others in the group (Lord, DeVader, and Alliger, 1986; Hogan, Curphy, and Hogan, 1994; Judge et al., 2002), as well as the degree to which an individual exerts influence on others (Hollander, 1964).

Contextual leadership should not be viewed as either hierarchal or shared. Instead, research suggests that teams engaged in a combination of both hierarchal and shared forms of leadership have the best outcomes (Pearce and Sims, 2002; Pearce, 2004; Ensley, Hmielski, and Pearce, 2006). Understanding ways in which more traditional and hierarchical leadership may be used in conjunction with more participative, shared, or otherwise emergent forms of leadership is particularly relevant for effective leadership of science teams and groups. For example, based on extensive, repeated interviews, Hackett (2005) found that the directors of successful microbiology laboratories at elite research universities used and valued both directive, hierarchical leadership and shared, participative leadership styles. It is also important to understand how shifts in leadership hierarchies occur in science teams and groups and how best to manage these shifts, depending on the stage of the research project or the expertise needed at different times.

RESEARCH FINDINGS ON TEAM LEADERSHIP

The general leadership theories delineated in the previous section have useful, but only indirect, implications for team effectiveness (Kozlowski and Ilgen, 2006). In part, this is because they focus on a general set of behaviors that are broadly applicable across a wide variety of situations, tasks, and teams. They neglect unique aspects of specific team tasks and processes and the dynamic processes by which team members develop, meld, and synchronize their knowledge, skills, and effort to be effective as a team (Kozlowski et al., 2009).

Leadership and Key Team Processes

As discussed in Chapter 3, team processes have been shown to be connected to team effectiveness, and existing research demonstrates that leadership can influence several of these team processes: *team mental models, team climate, psychological safety, team cohesion, team efficacy,* and *team conflict.* Leader behaviors that can influence each of these behaviors in ways that enhance team effectiveness are described below and summarized in Table 6-1.

Several leader behaviors can influence the development of *team mental models.* Marks, Zaccaro, and Mathieu (2000) found that when leaders provided pre-briefs describing appropriate strategies for carrying out team tasks, there were positive effects on team mental models, as well as team processes and performance. Other research has linked leader pre-briefs/

TABLE 6-1 Team Processes That Are Influenced by Leader Behaviors

Process	Leadership Behaviors That Influence the Process
Team Mental Models	• Providing pre-briefs describing appropriate strategies for carrying out team tasks and other planning strategies • Conducting debriefs and providing feedback
Team Climate	• Defining the mission, goals, and instrumentalities for teams • Considering effects on team climate of emphasis in communications to team members
Psychological Safety	• Coaching • Reducing power differentials • Encouraging inclusion
Team Cohesion	• Explicitly defining social structure • Promoting open communications • Modeling self-disclosure
Team Efficacy	• Creating mastery experiences that enable team members to build individual self-efficacy, then shifting the focus of team members toward the team • Providing task direction and socioemotional support
Team Conflict	• Anticipating conflict in advance and guiding team members through the process of resolving conflict by establishing cooperative norms, charters, or other structures (*preemptive approach*) • Guiding team members in working through conflicts, employing the following strategies: specifying the nature of the disagreement and encouraging team members to develop solutions to the problem, and fostering willingness to accept differences of opinion, openness, flexibility, and compromise (*reactive approach*)

discussions of planning strategies and debriefs/feedback to the development of team mental models (Smith-Jentsch et al., 1998; Stout et al., 1999).

Leadership can have a significant influence on *team climate*. Leader practices that define the mission, goals, and instrumentalities for teams can shape team climate (James and Jones, 1974), as do communications from team leaders, particularly in terms of what leaders emphasize to team members (Kozlowski and Doherty, 1989; Zohar, 2000, 2002; Zohar and Luria, 2004; Schaubroeck et al., 2012).

Psychological safety is a facet of team climate. Team leaders can foster a climate of psychological safety through coaching, reducing power differentials, and encouraging inclusion (Edmondson, Bohmer, and Pisano, 2001; Edmondson, 2003; Nembhard and Edmondson, 2006).

While research on the antecedents of *team cohesion* is limited, theory suggests that developmental efforts by team leaders (e.g., Kozlowski et al., 1996, 2009) are likely to have a strong influence on the team's formation and maintenance. Newcomers to teams tend to "respond positively to leader efforts to convey social knowledge, promote inclusion, and communicate acceptance" (Kozlowski et al., 1996, p. 269, citing Major and Kozlowski, 1991). Kozlowski and colleagues (1996) proposed that several leader behaviors therefore promote the development of team cohesion, including explicitly defining social structure, promoting open communications, and modeling self-disclosure.

Kozlowski and Ilgen (2006) identified several leadership behaviors that can influence the development of *team efficacy*. One such behavior is creating mastery experiences that enable team members to build individual self-efficacy, and then shifting the focus of team members toward the team's efficacy. Leadership efforts related to task direction and socio-emotional support have also been found to predict team efficacy (Chen and Bliese, 2002, as cited in Kozlowski and Ilgen, 2006).

As discussed in Chapter 3, *team conflict*, particularly within diverse teams such as interdisciplinary or transdisciplinary science teams, may be inevitable. Leaders can minimize the harmful effects of conflict on team effectiveness by actively employing conflict management strategies. Marks, Mathieu, and Zaccaro (2001) identified two approaches to conflict management: preemptive and reactive. Preemptive approaches involve anticipating conflict in advance and guiding team members through the process of resolving conflict by establishing cooperative norms, charters, or other structures. In a study of 32 graduate student teams, Mathieu and Rapp (2009) found that the quality of team charters was related to the quality of the teams' performance. Reactive approaches involve guiding team members in working through conflicts, employing the following strategies: specifying the nature of the disagreement and encouraging team members to develop solutions to the problem, and fostering willingness to accept differences of opinion, openness, flexibility, and compromise (Kozlowski and Ilgen, 2006).

Based on their analysis of in-depth interviews with members of successful and unsuccessful science teams and larger groups, and building on an earlier guide to team science (Bennett, Gadlin, and Levine-Finley, 2010), Bennett and Gadlin (2012) proposed the use of pre-emptive approaches to manage conflict. Specifically, they suggested that team leaders and members develop explicit collaborative agreements at the beginning of a new research project, articulating how decisions will be made, how data will be shared, how authorship of publications will be handled, and other matters. The process of developing such plans requires the members to discuss and reach agreement on potentially divisive issues in advance, building trust within the team.

Leadership as a Dynamic Process

Team leadership involves the ability to direct and coordinate the activities of team members; assess team performance; assign tasks; develop team knowledge, skills, and abilities; motivate team members; plan and organize; and establish a positive climate (Salas, Sims, and Burke, 2005). This is consistent with research that proposes a functional approach to understanding team leadership structures and processes (Morgeson, DeRue, and Peterson, 2010), conceptualizing effectiveness in terms of team needs, satisfaction, and goal accomplishment (Kozlowski and Ilgen, 2006).

This functional approach treats team leadership as a dynamic process necessitating adaptive changes in leader behavior, as opposed to treating it as a fixed set of static and universal behavioral dimensions. This implies that leaders must strive to be aware of the key contingencies that necessitate shifts in leadership functions, and they must work to develop the underlying skills needed to help the team maintain fit with its task environment and resolve challenges. Dynamic leadership is a process, not a destination; in other words, dynamic leaders recognize that they must always continue to adapt their behavior to best meet the changing needs of evolving projects. Given the dynamic nature of scientific research, leaders of science teams and groups may be more successful if they adopt a dynamic or functional leadership approach, are psychologically agile, and can use appropriate and varied modes of communication to engage with people from multiple generations, backgrounds, and disciplines.

Researchers at the Center for Creative Leadership proposed an approach that might hold promise for effectively incorporating both hierarchical and shared forms of leadership as is necessary in interdependent science teams (Drath et al., 2008). They proposed that setting *direction*, creating *alignment*, and building *commitment* is essential among people engaged in shared work, and argued that any action that enables these three elements to occur is a source of leadership. This source could be an individual, a collection of individuals, the task itself, or the external environment. An advantage of this approach is that rather than offering a lengthy list of various leadership functions and behaviors (or competencies), the focus is on just the three core leadership tasks: setting direction, creating alignment, and building commitment.

These core leadership tasks are relevant to teams and can be used as a way to understand the dynamic nature of team processes. For example, Kozlowski and Ilgen (2006) proposed that team effectiveness occurs when team processes are aligned with environmentally driven tasks. The core leadership task of creating alignment is consistent with this dynamic conceptualization of team effectiveness. In this sense, team leadership involves all processes that serve to improve team effectiveness. This type of leadership generally evolves throughout the life cycle of a team as the necessary tasks at hand are constantly changing.

Dynamic models of team leadership have two primary foci centered on *task cycles or episodes*, and the process of *team skill acquisition and development*. By harnessing cyclic variations in team task cycles to the regulatory processes of goal setting, monitoring/intervention, diagnosis, and feedback, the leader is able to guide team members in the development of targeted knowledge and skills—the cognitive, motivational/affective, and behavioral capabilities that contribute to team effectiveness. There is research evidence in support of this approach to team leadership from a meta-analysis of 131 effects relating team leadership to team performance, which found that team performance outcomes were associated with both task- and person-focused leadership (Burke et al., 2006). Specifically, Burke et al. (2006) found that task-focused leadership had a moderate positive effect on perceived team effectiveness ($r = .33$) and team productivity/quantity ($r = .20$), while person-focused leadership had almost no effect on perceived team effectiveness ($r = .036$), a moderate positive effect on team productivity/quantity ($r = .28$), and a larger positive effect on team learning ($r = .56$). Importantly, task interdependence was also shown to be a significant moderator in that leadership had a larger effect when task interdependence was high. The results of this research suggest that leadership in teams influences team performance outcomes by shaping the way team members work with core tasks, and by attending to the socio-emotional needs of the team.

A theory of dynamic team leadership, developed by Kozlowski and colleagues (Kozlowski et al., 2009), elaborates on the role of the formal leader in the team development process in helping the team move from relatively novice to expert status and beyond while building adaptive capabilities in the team. In these latter stages of team development, the team takes on more responsibility for its learning, leadership, and performance. In this manner, vertical and shared leadership operate sequentially with a formal leader helping the team prepare itself to take on the core functions of leadership and learning. Thus, building adaptive team capabilities or collective leadership capacity (Day, Gronn, and Salas, 2004) is an important team leadership challenge.

Tannenbaum and colleagues (2012) observed that the evolving drive for collaborative leadership reflects the changing nature of teams and the environments in which they operate. As team or larger group size increases, it becomes necessary for leaders to distribute certain leadership tasks, empower team members for more self-management, and create good learning opportunities for the members.

Current research suggests that team empowerment is facilitated by supportive organizational structures (Hempel, Zhang, and Han, 2012); team-based human resource policies for training, development, and rewards (Adler and Chen, 2011); and team-based and external reinforcing leaders (Kirkman and Rosen, 1999). Chen and Tesluk (2012) identified individual-level, team-level, and organizational-level antecedents to team empowerment. At the individual level, self-view, degree of self-efficacy, and need for achievement; job characteristics (such as level of ambiguity and unit size); and the quality of relationships with supervisors and coworkers influence team empowerment. At the team level, leadership behaviors, team climate, and team work characteristics can influence team empowerment. At the organizational level, organizational climate and human resource management practices such as employee development systems and team-based rewards and training were identified as possible antecedents to team empowerment (Chen and Tesluk, 2012).

Finally, the goal-directed activities of team task performance are cyclical in nature and constantly changing (Marks, Mathieu, and Zaccaro, 2001). This episodic perspective on team tasks distinguishes between *action* and *transition* phases of team performance, with the former focusing on task engagement and the latter on task preparation and follow-up reflection. This has important leadership implications. Specifically, there are certain processes or actions that are targeted at managing the team transition phase (e.g., mission analysis, goal specification, strategy formulation and planning), other actions targeted for the action phase (e.g., monitoring progress, systems monitoring, team monitoring and backup, coordination), and actions that are relevant for both transition and action phases (e.g., conflict management, motivating and confidence building, affect management). Dynamic models of team leadership can be conceptualized in contingency or contextual leadership terms, given that different actions or leadership functions are required in different phases of team performance. Consonant with this perspective, a recent study has proposed a model of transdisciplinary team-based research encompassing four distinct phases (Hall et al., 2012b).

Leadership and Team Faultlines

One area of research that is highly relevant to team leadership for effective team functioning is the topic of *faultlines*. As discussed in Chapter 4, faultlines are defined as boundaries that develop between subgroups within teams that detract from their overall effectiveness. Because faultlines escalate group conflict (Thatcher and Patel, 2012), their management, viewed within the construct of shared leadership, is essential for well-functioning teams. On the flip side, team conflict may also increase innovation by redirecting energy toward creating new ideas.

A strategy that leaders can use to mitigate subgroup conflict and strive instead toward innovation is to build superordinate team identification and superordinate goals (Bezrukova et al., 2009; Jehn and Bezrukova, 2010; Rico et al., 2012). Team identification and the strength of members' attachment to the group may bind members together into a powerful psychological entity (Ashforth and Mael, 1989; Chao and Moon, 2005; Van der Vegt and Bunderson, 2005). Empirical research has demonstrated better performance of faultline groups when team identification is high (Bezrukova et al., 2009). Another way leaders might reinforce superordinate team identification is by establishing common goals, norms, or cultural values. Cultural misalignment between subgroup values and those of the larger business unit has negative implications for performance (Bezrukova et al., 2012). Multicultural teams may be particularly vulnerable to the development of team faultlines. Fussell and Setlock (2012) discussed types of cultural variation and the effects on teamwork, and offered several strategies for overcoming challenges presented to leaders of culturally diverse teams, including offering culture-specific and diversity awareness training for team members, developing team interaction strategies to address particular cultural issues (such as providing an anonymous way to make contributions to team discussions when some members of the team are from a culture that discourages public disagreement with leaders), and using appropriate collaboration tools.

Another approach to mitigating conflict betweeen subgroups is to create a cross-cutting strategy such as a reward system or task role assignment that cuts across the larger group (Homan et al., 2008; Rico et al., 2012). For example, in a science team or larger group, engineers and scientists may be grouped together to work on different aspects of a prototype. The cross-cutting identification with the shared task would be expected to decrease bias and contribute to productive communication by reducing psychological distance between subgroups of engineers and scientists.

Finding common ground is yet another strategy that team leaders can use to leverage external conflict to make faultlines less salient. This approach unites the team to "fight" against common "enemies" outside the team (Tajfel, 1982; Brewer, 1999). In this way, the team members can perceive higher levels of team efficacy, autonomy, and relatedness, leading to increased team motivation and self-regulation (Ommundsen, Lemyre, and Abrahamsen (2010).

Intergroup Leadership

One area of research on leadership in business and government that may be relevant to leading science teams and larger group involves *intergroup leadership*. As Pittinsky and Simon (2007) discussed, leaders can encounter challenges in their efforts to foster positive relationships among subgroups of followers or constituents. Behaviors that foster subgroup or team cohesiveness can positively impact outcomes within the subgroup or team, but at a cost to relationships with other subgroups or teams, which can ultimately have a negative impact on outcomes of both the subgroups or teams and the larger business or governmental organization. This is similar to the challenge of leading multiteam systems discussed later in this chapter. Pittinsky and Simon (2007) discuss five leadership strategies for promoting positive intergroup relations: (1) encouraging contact between groups, (2) actively managing resources and interdependencies, (3) promoting superordinate identities, (4) promoting dual identities, and (5) promoting positive intergroup attitudes. Hogg, van Knippenberg, and Rast (2012) also discussed the importance of intergroup leadership and identify the leader's ability to promote an "intergroup relational identity" (p. 233) as critical to the development of positive intergroup relationships.

RESEARCH FINDINGS ON TEAM SCIENCE LEADERSHIP

In this section, we focus on the existing literature on science teams and larger groups and discuss the leadership challenges.

Models of Team Science Leadership

Because science teams and larger groups share many features with teams and groups in other contexts, their leaders can enhance effectiveness partly by facilitating the team processes shown to enhance effectiveness in other contexts, as shown in Table 6-1 above. Research and theory conducted in science contexts also suggest that leader behaviors to foster these processes will enhance effectiveness. For example, B. Gray (2008) proposed that transdisciplinary teams require leadership that creates a shared mental model or mindset among team members (i.e., cognitive tasks; see also O'Donnell and

Derry, 2005); attends to the basic structural needs of the team in terms of managing coordination and information exchange within the team and between the team and external actors (i.e., structural tasks); and also focuses on developing effective process dynamics within the team (i.e., procedural tasks).

B. Gray's (2008) view of collaborative team science leadership is conceptually very similar to shared leadership, discussed earlier. It may be tempting therefore to conclude that effective leadership in science teams can best be accomplished by facilitating collaborative and shared leadership processes; however, this conclusion may be both premature and overly simplistic. As noted above, Hackett (2005) found that the directors of successful microbiology laboratories at elite research universities used and valued both directive, hierarchical leadership and shared, participative leadership styles. Some of these science leaders adopted permissive, participative leadership styles, allowing students and colleagues autonomy to learn and develop their own approaches, while others were more forceful in their direction and follow more sharply drawn lines of inquiry. This apparent leadership paradox is consistent with the notion that there is no one best way to lead in terms of enhancing team effectiveness. Hackett (2005) proposed that the different leadership styles reflected each director's multiple roles as a scientist, leader, teacher, and mentor. Spending time in the laboratory may give a director greater control over technologies and subordinate scientists, but less time for writing the proposals, papers, and reviews that sustain the laboratory's funding and its identity within the larger scientific community. Over time, many of the directors had lost their cutting-edge scientific skills and become more reliant on the work of their followers, creating new tensions of leadership.

The research suggests that team science leaders would benefit from developing skills and behaviors that would allow them to practice directive as well as more participative, collaborative, or shared styles of leadership depending on team needs. This is consistent with the dynamic leadership processes described in the previous section.

Similar to studies in other contexts showing a relationship between leader behaviors, team processes, and team effectiveness, a study of academic science teams in Europe found significant positive relationships between supervisory behavior, group climate (a team process), and research productivity (Knorr et al., 1979). Supervisory quality was measured by surveys of followers' satisfaction, including survey items related to the supervisor's planning functions (e.g., satisfaction with the quality of research program, satisfaction with personnel policies) and integrative functions (e.g., satisfaction with group climate, feeling of attachment to the research unit). Within the overall positive relationship between supervisory quality and group climate, ratings of the supervisors' planning and integrative functions were the most important intervening variables.

One practical way to deal with the complexities of leading science teams or groups is through engaging the members to collectively develop a team charter, which provides a written agreement for task accomplishment and teamwork and has been shown to enhance effectiveness in teams outside of science (Mathieu and Rapp, 2009).

Emerging Team Science Models and Leadership Implications

The two models of team science described in Chapter 3 incorporate many of the leadership concepts discussed in this chapter, highlighting the potential value of professional development for team science leaders.

In their integrative capacity model, Salazar and colleagues (2012) proposed that leaders of interdisciplinary or transdisciplinary teams or larger groups can build the capacity for deep knowledge integration (one of the key features introduced in Chapter 1) through several leadership styles and behaviors. For example, leaders who use an *empowering leadership style* can enhance the use of the team's intellectual resources (Kumpfer et al., 1993). This facilitates equal access to dialogue that is often hindered by status and power differences (Ridgeway, 1991; Bacharach, Bamberger, and Mundell, 1993). *Building consensus* through team developmental strategies such as experiential learning and appreciative inquiry, another leadership technique, can help to develop agreement around goals and problem definition, ultimately facilitating integrative knowledge creation (Stokols, 2006). Leaders who *listen for places where clarification might be needed* are best placed to communicate knowledge across geographic boundaries (Olson and Olson, 2000). Finally, *conflict management* (i.e., minimizing team divisions, as in managing the faultiness discussed above) and *affect management* (i.e., the facilitation of trust between team members) can serve as effective ways in which to foster collaboration and knowledge generation (Csikszentmihalyi, 1994; B. Gray, 2008; Salazar et al., 2012).

The integrative capacity model has important implications for research on team science leadership. The model's authors are currently conducting a study to determine how the development of a team's integrative capacity and subsequent knowledge outcomes are impacted by boundary-spanning leadership behaviors and interventions. The research has the potential to fill a vital gap within the literature by both developing measures of these constructs and empirically testing the theoretical propositions linking integrative capacity to the creation of new knowledge in multidisciplinary teams. The authors will measure the constructs and test their relationship to the theoretical propositions using a large-scale highly controlled quasi-experimental research design a sample of more than 40 interdisciplinary and transdisciplinary science teams across several U.S. universities.

The four-phase model proposed by Hall et al. (2012b) provides a roadmap to enhance the development, management, and evaluation of transdisciplinary research (see Box 3-2). It includes four relatively distinct phases: development, conceptualization, implementation, and translation and suggests the use of several tools to accomplish the goals of each phase, such as research networking tools in the development phase (see Chapter 4), the "Toolbox" seminars during the conceptualization phase (see Chapter 5), and conflict management tools during the implementation phase. This new model suggests that leaders can play a valuable role by providing the appropriate tools at each phase and working to ensure that team members use and learn from these tools.

Role of Scientific Expertise

Most leaders of science teams and larger groups are appointed or elected to these positions based on their scientific expertise (Bozeman and Boardman, 2013), and there is some evidence that subordinate scientists rate the quality of their leaders primarily in terms of such expertise (Knorr et al., 1979; Hackett, 2005). B. Gray (2008) suggested that relevant scientific expertise is critical to the leadership behaviors of managing meaning and visioning in transdisciplinary science teams or larger groups.

> Leaders manage meaning for others by introducing a mental map of desired goals and the methods for achieving them while at the same time promoting individual creativity. . . . In transdisciplinary research, the cognitive tasks of leadership largely consist of visioning and framing. . . . This visioning process is referred to as intellectual stimulation by transformational leadership researchers, and includes leader behaviors that promote divergent thinking, risk taking, and challenges to established methods. Transdisciplinary leaders need to be able to envision how various disciplines may overlap in constructive ways that could generate scientific breakthroughs and new understanding in a specific problem area. They themselves need to appreciate the value of such endeavors, be able to communicate their vision to potential collaborators, and construct a climate that fosters this collaboration (2008, pp. S125–S126).

Similarly, Bennett and Gadlin (2012) proposed that effective team science leaders are able to articulate the scientific project vision, both to the research community and the home institution, in a way that allows each team member to recognize his or her contributions. Some leaders of large scientific groups have called for creating a new position, the interdisciplinary executive scientist. This role would be filled by individuals who have both project management skills and deep expertise in at least one of the disciplinary areas involved in the interdisciplinary endeavor.[3]

Leadership of Multiteam Systems

A *multiteam system* is a complex system of teams created to accomplish goals too ambitious for any single team (Zaccaro and DeChurch, 2012). The system may consist of various types of teams and involve different leadership structures (Marks, Mathieu, and Zaccaro, 2001). In science, multiteam systems may be engaged in interdisciplinary or transdisciplinary research projects, which aim to deeply integrate knowledge from multiple disciplines and perspectives (one of the key features introduced in Chapter 1). Team leaders as well as members face the challenges emerging from this feature, as they may be unfamiliar with disciplines and perspectives included in their projects.

Of direct relevance to the seven key features that generate challenges for team science, some factors thought to be important in motivating different forms of multiteam leadership include the overall size of the multiteam system, the amount and kind of diversity, geographic dispersion, the level of interdependence among component teams, and power distribution among teams. More mature multiteam systems are reported to display greater levels of shared leadership than less mature multiteam systems, which makes sense given that shared leadership takes time to develop (DeRue, 2011). An example of this evolution, described further in Box 6-1, is the shared leadership within the large multiteam system of physicists, engineers, and computer scientists conducting research enabled by the Large Hadron Collider in Switzerland. The development of this shared leadership approach within what has been described as a "communitarian culture" in particle physics was born of necessity, because the funding level required for such large facilities precludes funding similar projects in multiple locations. In light of the growth of multiteam systems, other disciplines than particle physics might benefit from a similar philosophy and leadership approach.

In multiteam systems, leaders can engage participants in developing charters as a way to develop effective norms for between-team communication and leadership processes (Asencio et al., 2012). The process of creating a charter can also be used to identify a representative from each team who would participate in system-level leadership, help coordinate multiteam actions, and convey information across team boundaries.

To date there has been relatively little empirical research on leadership in multiteam systems. One study involved analyses of critical incidents in

---

[3] See https://www.teamsciencetoolkit.cancer.gov/Public/expertBlog.aspx?tid=4&rid=1838 [April 2015] for further discussion of this proposed position.

mission-critical multiteam environments, such as disaster relief systems (DeChurch et al., 2011). Based on the analysis, the authors identified a set of leadership behaviors that promoted positive team and interteam processes and enhanced performance of the multiteam systems. These behaviors included formulating overall strategy and coordinating the activities of the component teams. In a laboratory study examining leadership functions hypothesized to be important in synchronizing multiteam systems, DeChurch and Marks (2006) manipulated leader strategizing and coordinating and assessed their effects on functional leadership, interteam coordination, and the performance of the multiteam system. Results supported a multilevel (i.e., team and multiteam) model in which leader training positively influenced functional leadership, which in turn improved inter-team coordination, and ultimately resulted in improved performance of the multiteam system.

LEADERSHIP DEVELOPMENT FOR TEAM SCIENCE LEADERS

Leader and team member skills and knowledge are essential to foster effective team science. This includes both scientific knowledge and skills relevant to the research problem at hand and knowledge and skills to foster positive team or group processes that, in turn, enhance scientific effectiveness. The previous chapter discussed education and professional development for team members. Here we discuss approaches to developing the skills and knowledge required for effective leadership of science teams and larger groups.

Research conducted in contexts outside science has found that formal leadership development interventions can help leaders develop the capacity to foster positive team and organizational processes, thereby increasing team or organizational effectiveness (e.g., Avolio et al., 2009; Collins and Holton, 2004). For example, in a meta-analysis of research on leadership and performance, Avolio et al. (2009) found, across 37 leadership training and development interventions, a positive corrected effect size (d) of .60. The authors also analyzed the return on investments in the training and development interventions included in the study. They found that investments in the interventions with moderate to strong effects would yield positive returns in improved performance. For example, for a mid-level leader, the return on an investment in a development intervention with moderate effects ranged from 36 percent for online training to 169 percent for on-site training. As noted above, in their laboratory study of multiteam system leadership, DeChurch and Marks (2006) found that leader training positively influenced functional leadership, which in turn improved interteam coordination, thereby improving the performance of the multiteam system.

BOX 6-1
CERN:
An Example of Successful Multiteam System Leadership

On July 4, 2012, the European Organization for Nuclear Research, also known as CERN, in Geneva, Switzerland, announced the observation of a new subatomic particle consistent with the Higgs boson. Described as the "Holy Grail" of physics, the Higgs boson is important to fundamental understanding of the universe because it helps to explain why matter has mass. The CERN laboratory, founded in 1954, includes the Large Hadron Collider and detectors built specifically to study the Higgs mechanism. The observation of the Higgs boson was announced by two groups made up of thousands of physicists, engineers, computer scientists, and technicians from around the world (ATLAS Collaboration, 2012; CMS Collaboration, 2012). Research to date suggests that the unique organizational structures (Shrum, Genuth, and Chompalov, 2007) and culture (Traweek, 1988; Knorr-Cetina, 1999) of particle physics contributed to this scientific breakthrough.

Following World War II, as physics developed into an important research field, investigators developed increasingly large and powerful particle accelerators and detectors to measure the activity of the particles. Groups organized around detectors functioned as semi-autonomous units, linked to others by exchanges of information, students, postdoctoral fellows, and technical gossip (Traweek, 1988). At CERN, the two groups that discovered the Higgs boson are referred to as "experiments" and are named for the detectors that are the focus of their research—the Compact Muon Solenoid (CMS) and ATLAS detectors—located within the Large Hadron Collider. Each experiment is a very large group within the CERN system, and each is composed of multiple layers of groups and subteams. This organizational structure reflects DeChurch and Zacarro's (2013) model of a multiteam system—an organization made up of multiple teams that work toward different team goals, but share at least one system-level goal.

DeChurch and Zaccaro (2013) propose that multiteam systems must balance the tensions of confluent and countervailing forces to succeed. Confluent forces, such as coordination within and across teams, combine across teams and jointly enhance the performance of the entire system. Countervailing forces, in contrast, operate in contradictory ways within and across teams, detracting from the performance of the entire system. For example, team cohesion and strong feelings of unique team identity may enhance team-level performance but compromise information-sharing across teams.

The CMS experiment (Incandela, 2013) includes approximately 4,300 scientists, engineers, and technicians from 42 countries and 190 institutions. Participants work in hundreds of subteams organized in two major categories: service and physics. The service category includes, for instance, a computing team that manages more than 100,000 computers in 34 countries and an offline team that manages reconstruction and analysis software. These teams collect petabytes of information (22 in 2011 and 30 in 2012) for analysis, and oversee the networking and computational resources to allow distributed access, called the grid. The physics category includes multiple groups, such as the Higgs group of approximately 700 physicists organized into five subteams (Incandela, 2013).

Both egalitarian and hierarchical, the experiment is led by consensus among physicists motivated both by common interests and by formal goals and decisions established by CERN and experiment leaders.* At the top level are a board with representatives of all of the collaborating national research institutions and an elected spokesperson who is the executive head of the experiment. Countervailing forces sometimes emerge from strong identification with a subteam or subgroup, usually because an overly ambitious subteam leader has difficulty with collaborative science. To address this, top leaders rotate subteam leaders every 2 years, often appoint two co-leaders, and, if there is potential danger to the entire experiment (the system level), they may intervene as a last resort to replace a problematic subteam leader. Countervailing forces are also dampened by the general approach of drawing subteam leaders from within the team. If they demonstrate excellent performance, they may have more influence when they return to the team, or they may be promoted—a possibility that may motivate them to maintain cohesion with other teams in pursuit of the higher-level goals of CMS.

To encourage confluent forces, CMS leaders engage in intense, ongoing, and transparent communications. They convene collaborationwide weekly meetings to discuss news, challenges, strategies, and plans. Almost all meetings are open to any participant (who may attend in person or by videoconference), and open discussion of any major shifts in strategy encourages all subteams to focus on systemwide goals.

At the same time, CERN leaders have worked to mitigate conflicts or countervailing forces between and within the two experiments. For example, in the early development of ATLAS, leaders used a slow, deliberative process to avoid conflicts between potential groups of participants. Through extensive consultation, they were able to break open established, and often competing, research groups and bring them into the project, as well as U.S. physicists who had been engaged in design and planning of the Superconducting Super-Collider (SSC), a project that was stopped by the U.S. Congress in 1993.

Particle physics has a unique "communitarian" culture, where verbal communication is of great importance and people meet frequently at large and small conferences and quickly disseminate information to each other (Knorr-Cetina, 1999). This culture encourages scientists to work for the common good. For example, the two papers announcing the discovery of the Higgs boson were authored by the "ATLAS Experiment" and the "CMS Experiment." An online appendix listed the 2,891 co-authors of the CMS paper, in alphabetical order, including all who contributed to any part of designing, building, operating, or analyzing data from the experiment. These publications reflect the established rule that any results are owned by the collaboration. Individuals cannot publish results before going through the regular process of review and approval inside the experiment, with input from the CERN publications committee. This internal review process is so thorough that journals trust the outcome with little further review—a practical solution since most of those with the technical expertise to serve as journal peer reviewers are affiliated with the experiments.

---

*Because most funding for CERN experiments is controlled by member institutions and nations rather than CERN directly, laboratory leaders rely heavily on consensus building to achieve their goals (Hofer et al., 2008).

Leadership development trajectories are influenced not only by formal training and leadership development programs, but also by experience in leadership positions. Day (2010) noted that deliberate practice is a very important component of leadership development, as is fostering a sense of identity as a leader, which can lead to greater interesting in learning about leadership and improving leadership skills (see Day, Sin, and Chen, 2004; Day and Harrison, 2007; Day, 2011; Day and Sin, 2011). In addition to the mechanisms of formal training programs and experiential learning, self-directed learning or self-development can play an important role in leadership development (see Boyce, Zaccaro, and Wisecarver [2010] for an examination of leaders' propensity for self-development). Formal leadership training interventions may work to improve leadership styles and behaviors partly by fostering participants' sense of identity as a leader, and thus supporting experiential and self-directed learning.

The scientific community has begun to recognize the potential benefit of formal professional development for team or group leaders. Efforts are under way to extend and translate the leadership research to science contexts, as briefly described in the examples below.

Science Executive Education

This program funded by the National Science Foundation (NSF) is designed to address the fact that science executives who manage science enterprises often learn on-the-job through trial and error, usually without benefit of knowledge from organization science that might help them. As is the case for business executives, science executives need expertise in organizational governance, innovation management, resource provisioning, workforce development, turnover reduction, process improvement, and strategic leadership. However, for important contextual reasons, such as the fact that the business focus is on competitive industries rather than the pre-competitive world of basic research, business education models usually cannot be directly applied to science. Science executives increasingly have to balance long-term versus short-term goals, temporary projects versus permanent organizations, planning versus spontaneous action, and standardization versus fluid technical innovation. Hence, the lack of science executive expertise is regarded as a "rate-limiter" to moving toward greater coordination and collaboration.

In response to this need, the Science Executive Education Program was developed, drawing on research on interorganizational governance, virtual teams, distributed team collaboration, and innovation management involving organizational learning and memory. Extending project management to entrepreneurial leadership is at the center of science executive education (Cummings and Keisler, 2007, 2011; Karasti, Baker, and Millerant, 2010; Claggett and Berente 2012; Rubleske and Berente, 2012). Science executive education focuses on four main areas: matching sources and uses for funds over time, explaining the "value-added" of centers to various constituencies, improving hiring and retention of key employees, and better handling of the "socio" in socio-technical systems.

Project Science Workshops

This program, which has been in existence for 11 years, aims to develop project management skills for leaders of large scientific research projects. Developed by astronomer Gary Sanders with support from NSF, the annual workshop uses didactic presentations and case studies to cover a range of project management challenges, including design of complex projects and the tools needed for their management.[4] Topics at the workshop have included large-scale collaborative science; building scientific structure and partnerships; and selection, governance, and management of unique large-scale research facilities. The 2012 workshop attracted scientists from a wide range of large projects, such as the Blue Waters supercomputer at the University of Illinois at Urbana–Champaign, the Summit Station Greenland facilities, the iPlant collaborative focused on creating cyber infrastructure and tools for plant biology, and the interdisciplinary team creating the Thirty Meter Telescope in Pasadena, California.

Leadership for Innovative Team Science (LITeS)

The Colorado Clinical and Translational Sciences Institute (CCTSI) developed the LITeS Program in 2008 to strengthen participants' leadership, to foster team science through the establishment of a network of researchers who can support one another, and to increase opportunities for researchers to collaborate across disciplines. The program is provided annually to a cohort of both senior and developing leaders working in clinical and translational research at the University of Colorado, and is structured as a full-year experience that includes participation in small-group projects and four workshops covering a variety of topics relevant to science team leadership, as well as individual feedback and coaching (Colorado Clinical and Translational Sciences Institute, 2014). The program description on the institute's website (Colorado Clinical and Translational Sciences Institute, 2014, p. 6) states that the LITeS Program "is designed to address three major domains for leadership: (1) knowledge of individual leadership styles and behaviors; (2) interpersonal and team skills for leading, managing, and working with people; and (3) process skills for increasing quality and efficiency in the work of academic leadership."

---

[4]More information is available at http://www.projectscience.org/ [April 2015].

ADDRESSING THE SEVEN FEATURES THAT
CREATE CHALLENGES FOR TEAM SCIENCE

The research findings on the general topic of leadership, team leadership, and science teams in particular address the challenges of team science in unique ways. The consistent theme from this research is that no single leadership style or behavior can be prescribed for effective leadership and management of science teams, but rather, a combination of approaches is required. This combination encompasses: shared and hierarchical leadership; contingency and dynamic leadership that recognize the cyclical and temporal needs of a team as it develops and evolves over time; goal alignment; and the management of faultlines within and between teams that manifest as conflict, including conflict that drives innovation. Moreover, emerging research suggests that leaders of science teams and larger groups can be helped to acquire leadership behaviors and management skills. In Table 6-2, we summarize how the research findings discussed in the previous section might be applied to address each of the team science features that can create communication and coordination challenges.

SUMMARY, CONCLUSION, AND RECOMMENDATION

Currently, most leaders of science teams and larger groups are appointed to their positions based solely on scientific expertise and lack formal leadership training. At the same time, an extensive body of research on organizational and team leadership in contexts outside of science has illuminated leadership styles and behaviors that foster positive interpersonal processes, thereby enhancing organizational and team performance. Extending and translating this research could inform the creation of research-based leadership development programs for leaders of science teams and groups. The committee expects that such programs would strengthen science team leaders' capacity to guide and facilitate the team processes, thereby enhancing team effectiveness.

CONCLUSION. *Fifty years of research on team and organizational leadership in contexts other than science provide a robust foundation of evidence to guide professional development for leaders of science teams and larger groups.*

RECOMMENDATION 3: Leadership researchers, universities, and leaders of team science projects should partner to translate and extend the leadership literature to create and evaluate science leadership development opportunities for team science leaders and funding agency program officers.

TABLE 6-2 Addressing Seven Features That Create Challenges for Team Science

Feature	Leadership Research Addressing the Challenges Emerging from the Feature
1. High Diversity of Membership	• Dynamic team leadership. Formal leader plays a key role in the development and prepares team to take on more responsibility over time (Kozlowski et al., 2009). • Adopting the view of *team performance cycles.* Understanding the four-phase model and how to approach each phase (Hall et al., 2012b). • Managing faultlines (Bezrukova et al., 2009).
2. Deep Knowledge Integration	• Setting *direction,* creating *alignment,* and building *commitment* (Drath et al., 2008).
3. Large Size	• Team empowerment for shared leadership (Tannenbaum et al., 2012).

4. Goal Misalignment with Other Teams	• *Direction, alignment,* and *commitment* (see #2 above).
	• Developing team charters.
	• Leadership training, developing integrative capacity (Salazar et al., 2012):
	  o Empowering leadership styles (Kumpfer et al., 1993).
	  o Building consensus (Stokols, 2006).
	  o Listening for places where clarification might be needed (Olson and Olson, 2000).
	  o Conflict and affect management (Csikszentmihalyi, 1994; B. Gray, 2008).

5. Permeable Team and Group Boundaries	• Contingency leadership and the four-phase model (Hall et al., 2012b). • Develop a shared mental model or mindset among team members (i.e., cognitive tasks); attend to basic structural needs of the team in terms of managing networks that develop among interdisciplinary and transdisciplinary teams (i.e., structural tasks); and focus on developing effective process dynamics within the team (i.e., procedural tasks) (B. Gray, 2008). • Leader and team member behaviors oriented toward bridging disparate networks to facilitate knowledge generation and integrative capacity. See also "leadership training" in #4 (Salazar et al., 2012).
6. Geographic Dispersion	• See discussion in Chapter 7.
7. High Task Interdependence	• Task-focused leadership. Leadership is important when task interdependence is high. Leadership can shape the way team members work on core tasks and should attend to the socioemotional needs of the team (Burke et al., 2006).

Part III

The Institutional and Organizational Level

7

Supporting Virtual Collaboration

As science attempts to answer bigger and bigger questions, it is more and more likely that the people participating in the effort together reside in different locations, institutions, and even countries. As noted in Chapter 1, scientific publications are increasingly written by teams and larger groups across institutional boundaries (Jones, Wuchty, and Uzzi, 2008). Geographic dispersion is one of the seven features that can create challenges for team science, particularly with communication and coordination. This chapter begins by delineating these challenges. We then describe, in turn, the findings of the literature on how these challenges are met by the individual members of the distributed team or larger group, the team or group leaders, and the organizations that wish to support distance collaborations.

Because many of the disadvantages that arise from being distant from one's colleagues can be mitigated by various kinds of technologies, we next describe the suite of technologies available to support distance science. We then summarize how technology addresses some of the challenges of being geographically distributed. This chapter focuses on addressing a single feature of team science that creates challenges. Therefore, we do not include a separate discussion of the seven features that create challenges for team science as we do in Chapters 4 through 6. The chapter ends with conclusions and recommendations.

The chapter draws on many rich case studies of large groups and organizations[1] composed of geographically distributed scientists and other pro-

---

[1] As noted in Chapter 1, an organization typically incorporates a differentiated division of labor and an integrated structure to coordinate the work of the individuals and teams within it.

fessionals, which are supplemented by focused experiments and large-scale surveys and analyses of public records. For example, starting in the 1990s in the United States, the National Science Foundation has sponsored the development of a new organizational form for scientific collaboration called the Collaboratory (Wulf, 1993; Finholt and Olson, 1997)—a laboratory without walls. In Europe, this movement is called eScience or eResearch (Jankowski, 2009). To address science problems that are increasingly large and complex, collaboratories combine experts from multiple universities. Thus, they are typically geographically distributed, encountering all the issues outlined in this chapter in addition to those discussed earlier. The Science of Collaboratories Database (Olson and Olson, 2014) lists more than 717 such collaboratories, mainly in science but also in the humanities. Many of the entries include information about the topic, the participants, the shared instruments (such as the Large Hadron Collider) if any, funding, and the type of collaboratory, based on a proposed typology.

SPECIAL CHALLENGES FOR GEOGRAPHICALLY DISPERSED LARGER GROUPS OR TEAMS

Challenges for geographically dispersed groups include members being blind and invisible to one another; time zone differences; differences across institutions, countries, and cultures; and uneven distribution of members across participating locations.

Being Blind and Invisible

People working with others at distant locations are both invisible to those colleagues (Bell and Kozlowski, 2002) and blind to their actions and situations. In addition, people working virtually with remote colleagues are often unaware of the detailed context of those colleagues' work (Martins, Gilson, and Maynard, 2004). Research has shown that face-to-face communication is a valuable contributor to team performance (Pentland, 2012). Without *explicit* communication (Olson and Olson, 2000) or opportunities for periodic in-person visits, remote others do not know what individuals are working on, what their roadblocks and challenges are, and how they can help or be helped (Cramton, 2001). Technology solutions such as those outlined later in this chapter can help provide group members with the awareness they need to collaborate effectively, but group members must use these tools for this to happen. In other words, people need to take extra effort to report to remote others what they are working on, what the open issues are, and in general what the current context of work is, using e-mail, videoconferencing, teleconferences, or other electronic media.

There are additional issues of awareness not about the details of work but about the higher-level context of work. For example, a manager might unwittingly schedule a meeting during a remote location's predicted blizzard, or, crossing country boundaries, during hours outside of their normal workweek (e.g., people in France typically work a 35-hour workweek, having Friday afternoon as part of the weekend). Conversations that include people at the same location may also include references to weather, politics, and sports familiar to the local participants, but not to those in remote locations (Haines, Olson, and Olson, 2013). Finally, people starting a virtual collaboration may have difficulty establishing a work norm, and individuals joining an existing virtual group may have difficulty learning and adhering to such a norm once it has been established.

Time Zone Differences

Scheduling meetings that include participants from around the world can be a challenge because of people working with collaborators in different time zones. Constraints on available meeting times can range from merely being an hour off to having no overlap in people's working days (Kirkman and Mathieu, 2005). These constraints can lead to inconveniences to group members, such as the need to calculate and document accurate times among collaborators. Alternatively, some group members may have to make compromises to their own schedules, such as meeting early in the morning before their typical workday begins, during lunch, or late in the evening (Massey, Montoya-Weiss, and Hung, 2003; Cummings, Espinosa, and Pickering, 2009). Such compromises are more often made by the "minority" group member (the one individual on the other side of the globe) and can result in resentment or burn-out.

Differences Across Institutions

Science groups increasingly cross university boundaries. Academic institutions have different teaching schedules (some schools are on the quarter system, some semester, some intensive 8-week sessions). Different institutions also have different interpretations of rules about use of human subjects or about who owns intellectual property (Cummings and Kiesler, 2005, 2007). In addition, academic institutions use different technologies.

Differences Across Countries

Crossing country boundaries can create challenges regarding laws and expectations about intellectual property. In particular, regulations about the use of scientific specimens can differ, especially in human medicine. Laws and expectations related to intellectual property differ not only in terms of ownership of discoveries, but also in terms of the use of ideas and writings of others, as expressed in different definitions of copyright and plagiarism (Snow et al., 1996). Expectations can also differ around protecting the privacy of human research subjects (e.g., by requiring individuals to sign informed consent forms), the use of data and software (with or without license), and how data are managed and shared.

Differences Across Cultures

Even more subtle than differences in laws and expectations about intellectual property are differences in unspoken norms of work, definitions of various terms, and work style expectations (Kirkman, Gibson, and Kim, 2012). For example, in the United States, organizational decisions are often made by a high-level steering group and then announced so that others will buy in. In Japan and India, the decision-making process is much more consultative, as decisions are worked out in small groups to gain buy-in before being announced more ceremonially to the whole organization (Gibson and Gibbs, 2006). Subtle factors about conversational style also can differ. For example, in some cultures, the pauses in conversation are long, allowing time to think and honor what was just being said; in other cultures (in particular the United States), conversations progress at a rapid pace and people may "step on" each other's sentences and start to speak. A conversation including people from these two cultures can create impressions of disrespect on the one hand and assessment that the other has nothing to say on the other. Although beyond the scope of this report, there are many cultural differences when working across country boundaries, and these can have important effects on communication and ultimately effectiveness (Fussell and Setlock, 2012). Some very large, geographically distributed research organizations (e.g., CERN; see Box 6-1) provide support for these challenges, but other international groups are left to deal with these challenges on their own.

Uneven Distribution of Group Members Across Participating Locations

Often, members of geographically dispersed groups are not evenly distributed across all participating locations (O'Leary and Cummings, 2007). There is commonly a "headquarters" that involves the largest number of people, and satellites of one or two people included because of their special expertise. This is often referred to as the "hub and spoke" model. The culture and communication style of the headquarters typically dominate, and the group members at remote locations may experience lower status and less power, while their needs and progress are invisible to others (Koehne, Shih, and Olson, 2012).

Power and attention are more evenly distributed if each location has a critical mass of people, although this presents its own challenges. As noted in Chapter 5, Polzer et al. (2006) found that having subgroups based on geography was associated with higher conflict and lower trust. In particular, conflict was highest and trust was lowest when there were two co-located subgroups (e.g., half of the group members were in one country and half in another). Similarly, O'Leary and Mortensen (2010) found that when there is a critical mass of participants at several locations, the individuals have a tendency to form "in-groups" and "out-groups," with a tendency to disfavor and even disparage the out-groups.

INDIVIDUAL CHARACTERISTICS TO MEET THE CHALLENGES

As discussed in Chapter 4, individuals with social skills, such as those who score high on personality inventories as extroverts, are more likely to easily monitor and respond appropriately to actions and attitudes of others in their group or team (McCrae and Costa, 1999). Social skills are likely to be especially valuable in distributed groups, given that members need to communicate regularly and explicitly about the work being done.[2] An additional individual characteristic that may be valuable is being trustworthy (Forsyth, 2010). Trust is an important binder of any group or team, and engendering trust is especially important when members have infrequent contact with each other and few opportunities to directly interact face-to-face (Jarvenpaa, Knoll, and Leidner, 1998).

As discussed in detail later in this chapter, working in a distributed group involves communication and coordination through collaboration technologies ranging from e-mail and audio/videoconferencing to more sophisticated systems for scheduling time and sharing documents or data. Thus another salient member characteristic is technological readiness—a disposition to learn new technologies and to access training to make the learning easy. Also required at the individual level is the openness to explore new ways of working, in which one explicitly communicates actions that normally require no special thought (Blackburn, Furst, and Rosen, 2003). In addition, the individual must be willing to commit the time needed to learn the new technologies, both to get started and then to share best practices as the technology is adapted to the work.

Because remote collaborators cannot see and interact with each other directly and may have to overcome divisive boundaries, they often must

---

[2] As discussed in Chapter 5, individuals can be trained to develop social and interpersonal skills.

learn new habits of working, many of them through technologies. In addition to good e-mail habits (e.g., acknowledging receipt of e-mail even though there is no time at the present to respond fully), people have to learn to explicitly make their actions available to others so they are aware of progress and obstacles (Cramton, 2001).

LEADERSHIP STRATEGIES TO MEET THE CHALLENGES

There is growing evidence that effective leadership can help science groups and teams meet the challenges of collaborating across long distances. For example, Hoch and Kozlowski (2014) conducted a study of 101 virtual teams and found that when teams were more virtual in nature, traditional, hierarchical leadership was not significantly related to team performance, whereas shared leadership (discussed in Chapter 6) was significantly related to performance. This result was expected because the lack of face-to-face contact and often asynchronous nature of electronic communication makes it more difficult for team leaders to directly motivate members and manage team dynamics. The authors also found that for these teams, structural supports were more related to team performance than hierarchical leadership. Structural supports provide stability and reduce ambiguity in ways that may compensate for the uncertainty that characterizes virtual environments. Such supports include providing fair and transparent rewards for virtual teamwork and maintaining ongoing, transparent communications while managing information flow. These and other leadership strategies that can help increase the effectiveness of virtual science teams are discussed below.

Leading Virtual Groups or Teams

One of the important leadership activities for distributed groups occurs in meetings. Meetings present a challenge because of the unreliability of audio/videoconferencing, and the lack of cues about who would like to speak next or people's reactions to what is being said. The leader must explicitly solicit commentary and contributions from everyone, even polling individuals across all locations (Duarte and Snyder, 1999). This ensures not only that needed information and opinions are heard, but also that those at the smaller, distant locations feel respected for being asked. Also, when scheduling meetings among people who reside in disparate time zones, it is important that the leader fairly distribute the inconvenience of working outside of regular work hours to participate in the real-time meeting (Tang et al., 2011).

Also, the leader must be proactive in finding out what team or group members are doing (Duarte and Snyder, 1999). In a co-located setting, this is done by informally walking the hallways. In a distributed science group or team, it requires regular contact with all members. Frequent contacts, by e-mail instant messaging, voice, or video, are critical to supplement more formal scientific or technical progress reports. This contact also helps members know that they are valued members of the collaboration.

Managing Group or Team Dynamics

Common Experiences

The experience bases of individuals from different locations are likely to differ more greatly than the experience bases of individuals who are co-located. As discussed in Chapter 3, shared experience facilitates the development of two interpersonal processes that have been shown to enhance team performance—shared mental models (shared understanding of goals, tasks, and responsibilities) and transactive memory (knowledge of each team member's unique expertise (Kozlowski and Ilgen, 2006). In addition, team members' direct interactions shape team climate (shared understanding of strategic imperatives)—another process shown to improve team effectiveness. As such, virtual teams and groups are more likely to be successful if they engage in activities designed to overcome the lack of opportunities for shared experience, focusing, for example, on establishing common vocabularies and work style as explicit goals (Olson and Olson, 2014). This is especially important if the members come from different institutions and/or cultural backgrounds. Kick-off meetings are often used as a forum for members to explicitly assess habits and expectations, discuss differences, and agree on ways to resolve differences to increase chances for success (Duarte and Snyder, 1999).

Enhancing Readiness for Collaboration

To enhance readiness for collaboration, leaders can engage with members to foster intrinsic motivations, create extrinsic motivations, develop trust and respect, and thus improve group or team effectiveness. Individual members of a distributed group may have intrinsic (internal) motivation to work with the other members, either through personal ties or based on the realization that they need each other's expertise in order to succeed. Both of these behaviors generate respect; when people feel they are respected, they are more likely to be motivated to contribute (Olson and Olson, 2014). If these conditions do not hold, then the leader may need to create extrinsic (external) motivators, including group rewards and individual incentives that reflect how well the person contributed to the group (discussed further in Chapter 8).

Activities designed to foster trust and team or group self-efficacy—two other team processes shown to enhance effectiveness in non-science teams— may bolster the chance of a science team or group's success. First, because trust is slow to develop in a distributed group (with fewer occasions for people to learn how trustworthy others are and to become familiar with others' personal lives), leaders could provide exercises or activities for developing trust. For example, virtual chat sessions, in which people are encouraged to talk about their non-work lives and share things about them- selves that indicate vulnerability, have been shown to build trust (Zheng et al., 2002). Although such sessions can be valuable, the need to develop trust is one of the primary reasons that many teams conduct a face-to-face meeting of all participants at the outset of a project. Engaging the partici- pants in team professional development activities can also build teamwork and trust (see Chapter 5).

The second, related interpersonal process that helps ensure success is team or group self-efficacy, an attitude of "we can do it" (Carroll, Rosson, and Zhou, 2005; see Chapter 3 for further discussion). This attitude en- courages people to do extra work or find solutions when obstacles arise. Again, team-building exercises can help engender this attitude. As with trust, team self-efficacy enhances success in co-located as well as distributed teams, but when team members are distant, these processes are harder to establish and maintain.

Nature of the Work

When work is routine, such as on an auto assembly line, most people know what to do and what others are doing to coordinate their work. When work is complex, it is more challenging to keep track of what needs to be done and who is going to do which tasks. Collaborating at a distance is particularly difficult when the work is complex, as it is in team science (Olson and Olson, 2000). For example, in a study of 120 software and hardware development projects that were high in complexity, Cummings, Espinosa, and Pickering (2009) found that spatial boundaries (working across different cities) and temporal boundaries (working across time zones) were both associated with coordination delay. Coordination delay was defined in this study as the extent to which it took a long time to get a response from another member, member communication required frequent clarification, and members had to rework tasks.

One solution for managing complex work at a distance is to divide up tasks into modules so that most of the coordination and discussion happens among people who are co-located, essentially reducing the critical commu- nication required across locations (Herbsleb and Grinter, 1999). Because of the stresses of distance to awareness, communication, and coordination, the design of the work is critical (Malone and Crowston, 1994), and cognitive task analysis may aid in the distribution of work (see Chapter 4). If it is not possible to change the design of the work, then group or team members will be required to engage in extensive efforts to coordinate their research tasks.

HOW ORGANIZATIONS CAN SUPPORT VIRTUAL COLLABORATION

Geographically distributed science teams and larger groups are typically composed of members from separate organizations (e.g., universities). The culture and incentive structures of these organizations influence the collaborative readiness of groups or teams that cross its boundaries. An organization's culture sets the stage for the degree of competitiveness among, and status of, its members. The members within an organization work to act in ways that are aligned with reward structures. Misalignments, due to the incentive structure being individually focused versus team-focused or knowledge-driven versus product-driven, can have deleterious effects on its members' ability to successfully engage in team science.

In academics, disciplines vary in their competitiveness. For example, some scientists conducting research on Acquired Immune Deficiency Syndrome (AIDS), such as geneticists, immunologists, and pharmacists, may be intensely competitive because of the large amount of money and prestige associated with finding a cure. In another example, scientists in the Bio-Defense Center, a consortium of organizations in the northeastern United States funded by the National Institute of Allergy and Infectious Diseases, did not initially share their data because of fear of being "scooped" by someone publishing findings from their data before the data originator could do so. Coordination of distributed work is always easier when a scientific discipline or community has a culture of sharing and cooperation (Knorr-Cetina, 1999; Shrum, Genuth and Chompalov, 2007; Bos, 2008).

In projects requiring individual scientists to submit data to a shared repository, reward structures (e.g., based on use of the data by others) may be needed to motivate people to share their data (Bos, Olson, and Zimmerman, 2008). GenBank, a genetic sequence database of the National Institutes of Health (NIH), requires genomic data to be entered into the database as a precondition for publishing. The Alliance for Cellular Signaling worked with *Nature*, a highly prestigious journal, to develop a new process to review and publish a database of "Molecule Pages" (Li et al., 2002). These datasets are the standard format for the output of hard work by the scientists, but differ from traditional publications. *Nature* editors would then certify this review process when young professors came up for tenure with these kinds of publications. In 2010, there were 606 Molecule

Pages published, 88 under review, and 203 under preparation (see further discussion of authorship, promotion, and tenure in the following chapter).

Competition can also play a role in scientific research. Not only is it a great motivator, but also it is the most immediate source of corroboration and error correction. Creating parallel teams is common in particle physics. For example, as detailed earlier in Box 6-1, two separate teams built and operated different detectors at the Large Hadron Collider in order to find and examine the Higgs particle. These large international teams worked independently and announced their results simultaneously, yielding two broadly consistent sets of results that have been accepted with high confidence.

The leader of a distributed science group or team is often affected by decisions made at the organizational level, such as the university. For example, incentive structures are often dictated by the organization, and the culture of collaboration and/or competition is often strongly influenced by the entire organization or even profession. The organization may dictate the design of the research project or designate how many people are located at each site, which can in turn affect how interdependent the tasks are, with the consequent stresses on communication and coordination. The funding agency or organization ultimately determines the project budget, which in turn dictates how much money is available for technical capabilities and support. Although the leader can argue for the importance of technology suites, support, and training to facilitate remote collaboration, the keeper of the funds often makes the final allocation.

When multiple organizations are involved, as is often the case in long-distance collaborations, there are additional issues to work out. Explicit efforts to align research goals across institutions may delegate the institution-specific goals to a secondary level. Legal and financial issues may have to be negotiated, for example, to reconcile varying approaches to allocation of project funds in different countries. In large academic research projects, there are issues related to who gets credit for the results, not just the publications, but at the organizational level, as well as who gets credit for the funding award and who owns the intellectual property.

Although many organizations seek to foster flexibility and creativity through a flatter organizational hierarchy, this approach works best for co-located teams, where it is easier to communicate and share context and tacit information. For large, distributed groups, work goes more smoothly with at least some authority and designated roles and responsibilities (Hinds and McGrath, 2006; Shrum, Genuth, and Chompalov, 2007). One recent study found that leadership that is shared and provides structural supports (e.g., providing fair and transparent rewards for virtual teamwork, managing information flow) improved effectiveness in distributed teams (Hoch and Kozlowski, 2014).

Organizations and group leaders may benefit from the use of an online assessment tool called the Collaboration Success Wizard, see http://hana.ics. uci.edu/wizard/ [May 2015]. This tool asks the participants in a particular team science project to answer approximately 50 questions about the nature of the work, the motivations, the common ground, the management, and the technology needs/uses in the project. The respondent can ask for immediate feedback on where the team or group is strong, where vulnerabilities might lie, and, importantly, what to do about them. Following completion of the surveys, project leaders can obtain a summary report, again showing strengths, vulnerabilities, and what to do about them, because there are occasions when different individuals or subgroups may have different views about their work.

TECHNOLOGY TO SUPPORT VIRTUAL COLLABORATION

In this section, we first review the kinds of technologies that have been used to support distributed work, with different kinds of work benefiting from different constellations of technologies. The committee's framework follows closely that of Sarma, Redmiles, and van der Hoek (2010), categorizing technologies as communication tools, coordination tools, and information repositories, adding significant aspects of the computational environment (see Box 7-1). Although we refer to specific technologies, the point is not to recommend a specific current technology, because it will quickly be replaced with newer versions. We rather wish to emphasize the *types* of technology that are useful and why. We then present an analytic scheme to guide people in choosing the right constellation of technologies for their work.

Types of Collaboration Technologies

Communication Tools

E-mail and Texting E-mail is ubiquitous, and many experts have characterized it as the first successful collaboration technology (Sproull and Kiesler, 1991; Satzinger and Olfman, 1992; Grudin, 1994; Whittaker, Bellotti, and Moody, 2005). One of the cornerstones of its success is that today it is independent of the device or application used to send and receive it, and, with attachments, it is a way to share almost anything the recipient can read. As happens with other technologies, people also use it for managing time, reminding them of things to do, and keeping track of steps in a workflow (Mackay, 1989; Carley and Wendt, 1991; Whittaker and Sidner, 1996; Whittaker, Bellotti, and Moody, 2005).

BOX 7-1
Classification of Technologies to Support Distance Work

Communication Tools
    E-mail and texting
    Voice and videoconferencing
    Chat rooms, forums, blogs, and wikis
    Virtual worlds

Coordination Tools
    Shared calendars
    Awareness tools
    Meeting support
    Large visual displays
    Workflow and resource scheduling

Information Repositories
    Databases
    Shared files
    Laboratory notebook (online)

Computational Infrastructure
    System architecture
    The network
    Large-scale computational resources
    Human computation

SOURCE: Olson and Olson (2014). Reprinted with permission.

Instant Messaging (IM), sharing primarily simple text messages with another person or even a group, has made significant inroads into organizations. In some cases, it has replaced the use of e-mail, phone, and even face-to-face communication (Muller et al., 2003; Cameron and Webster, 2005). There is evidence that it is sometimes used for complex work discussions, not just simple back and forth about mundane issues (Isaacs et al., 2002). It is also used effectively for quick questions, scheduling, organizing social interactions, and keeping in touch with others (Nardi, Whittaker, and Bradner, 2000).

Except for e-mail attachments (which can include elaborate drawings, figures, and videoclips), the technologies listed above are text-based, even in the abbreviated world of texting. Text remains an impoverished medium compared to the tones and facial/body expressions possible in face-to-face communication.

Conferencing Tools: Voice and Video There are a myriad of opportunities to communicate beyond text in today's world, and many are used heavily. The telephone trumps text in being able to convey tone and to have immediacy of response. However, delays caused by technical interruptions of voice and video transmission are highly disruptive to conversational flow because of the importance of pauses in turn-taking in a conversation (Börner et al., 2010).

Many people have telephones from which they can teleconference, at least on a small scale. Organizations often provide services for larger-scale audio "bridges" for conference calls. Key to the smooth execution of these calls is whether the phones have "full-duplex" or "half-duplex" transmissions. Half-duplex lines are capable of transmitting only one direction at a time. Natural conversations often include "backchannels"—the "uh huh," "hmms," and other comments that convey whether the recipient is agreeing, understanding, or not; when using a half-duplex line, these responses are silenced. As a consequence, often the speaker will talk longer than necessary, not sure if the recipient has understood (Doherty-Sneedon et al., 1997). Additionally, conversational turn-taking is often signaled by an utterance from the one who wants to take the turn while the current speaker is speaking (Gibson and Gibbs, 2006). These are entirely cut out in a half-duplex line, creating awkward competitions for who will speak next.

Although tone of voice can add meaning to the words said, facial expressions and body language add another layer. In large meetings, video helps convey who is present without an explicit roll call, and by eye contact and expression, conveys who is paying attention. One can see not only the people but also the situation or context they are witnessing.

The richness of voice and video, however, can create barriers to people who are from different cultures. As noted earlier, the expected pause structures in conversation are different in the Western and Eastern cultures, often creating miscues. Because Westerners are used to a shorter pause structure than Easterners, they will dominate the conversation (Hinnant et al., 2012). Similarly, when video shows facial expressions and eye contact information, because those modes of expression are interpreted differently in different cultures, people again may make wrong attributions of interest and consent.

For greatest effectiveness, a video connection should be arranged to mimic a sense of physical presence. Eye contact and gaze awareness are key linguistic and social mediators of communication (Kendon, 1967; Argyle and Cook, 1976). In video, as in real life, people tend to focus on the face of the person with whom they are talking and attempt to make eye contact by looking at the eyes of the person. Unfortunately, to appear to make eye contact over video requires a person to look not at the projected eyes of the remote person but at the camera. Therefore, to convey eye contact, extra effort needs to be expended to move the video of the remote person as close to the camera as possible. Without this careful adjustment, meeting participants will appear as if they are glancing sideways or at the top of other participants' heads, both of which can be interpreted as disinterest (Grayson and Monk, 2003).

Conversations are often accompanied by gestures referring to an object, a document, data, or a visual image. Today, sophisticated tools, such as GoToMeeting, Google Hangout, and Skype screen-sharing allow a participant to share his or her computer desktop or a particular window with others, allowing them to control what they are looking at and the ability to focus attention by using the mouse/pointer.

Blogs, Forums, and Wikis   Longer conversations from larger numbers of people are usually accomplished through chat rooms, blogs, forums, and wikis. Chats are nearly real-time, whereas blogs, forums, and wikis have a longer time between contributions. When used for distributed science, all are typically restricted to a designated work group rather than being public.

The large groups of space physicists participating in the Upper Atmospheric Research Collaboratory and Space Physics and Aeronomy Research Collaboratory used chats extensively to converse during their "campaigns," periods when the sun's activity impacted the upper atmosphere. The automatically recorded chats allowed people to "read in" to the conversation (scrolling back and reading what had been happening), helping them "catch up" although their time zone differences prevented them from participating in "real time." The conversations were comparable to those held face-to-face (McDaniel, Olson, and Magee, 1996).

Wikis similarly are free-for-all conversations, but are even less structred in formatting. Forums are typically set up for discussion threads, whereas wikis can take any form whatsoever. The large groups of scientists participating in the Biomedical Informatics Research Collaboratory used wikis extensively to share test protocols, tips, frequently asked questions, announcements of the availability of new software tools, and articles of interest (Olson et al., 2008).

Virtual Worlds   Virtual worlds are graphical, 3-D representations of physical spaces and have drawn considerable attention from both industry and academia (Bainbridge, 2007). They allow a person to experience a realistic environment, usually through an avatar. Avatars can explore a space, manipulate objects, and, when networked together, interact with other people's avatars. The Meta-Institute for Computational Astrophysics is a collaboratory based exclusively in virtual worlds. The institute provides professional seminars, popular lectures, and other public outreach events in the game Second Life[3] (Djorgovski et al., 2010).

Such simulations of real worlds have been in common use for training in the military for a long time (Johnson and Valente, 2009). Although multiplayer games such as World of Warcraft[4] also allow for a wide range of playful interactions, Brown and Thomas (2006) speculated that real leadership skills might be learned in a game such as this because it involves extensive quests with a substantial numbers of players.

Coordination Support

A class of technologies exists to support collaborators in finding a time to work synchronously, and a second set of technologies supports coordination during their time together.

Shared Calendars   Although the original introduction of group calendars was met with resistance, many organizations have seen value in their use (Grudin, 1994; Grudin and Palen, 1995). Calendars support the coordination of meetings, finding a time when the important participants are available.

Calendars are also used as a tool to display and/or read availability. When colleagues do not respond to requests in their usual timely way, one can view their calendars to discover whether they are out of town or in a meeting. The information also allows for planning when to contact a person (e.g., an "ambush" after an in-person meeting in order to get a signature). Shared calendars can be particularly valuable for geographically dispersed colleagues who are in different time zones, reminding people of when the workdays overlap and where they do not.

Awareness Tools   Today, awareness information is conveyed in the status indicators of IM systems. With IM, the user has control over what status indicator to convey to others, but the feature comes at the cost of remembering to set it and actually setting it. The cost of receiving the status setting, however, is very low. Many IM clients list the person's chosen colleagues who agree to be monitored, and their status is typically listed in iconic form on the edge of the screen.

IM indicates the user's current state, from which others can infer whether she or he can be interrupted, but not specifically what they are

---

[3] For more information on Second Life, see http://www.secondlife.com [May 2015].

[4] For more information on World of Warcraft, see http://us.battle.net/wow/en/us.battle.net/wow/ [May 2015].

doing. In the domain of software engineering, a key form of advancement in science, where coordination of detailed efforts is of primary importance but the work nearly invisible, developers have created and widely adopted various system to "check out and check in" portions of the code they are working on. For example, Assembla[5] is a collection of tools to track open issues and who is working on them, plus a code repository where code is assigned to a person to work on, during which time others are locked from editing. These kinds of coordination tools are powerful, but not widely adapted to domains other than software engineering.

A more general system that notes what people have been or are working on in a shared document appears in Google Docs. The names of others who are currently editing the document are shown at the top of the document, and their cursors with their names in a flag are shown where they are working now. In addition, the list of past revisions and an indication of who did what (with authors' contributions highlighted in different colors) show what has been changed. These various symbols and colors provide awareness of others' efforts on a common document, useful if more than one person is working on the document either at the same time or asynchronously.

Meeting Support   Coordination support for meetings, whether they are face-to-face or remote, can be formal and informal. During the 1990s, developers and users tested Group Decision Support Systems, in which participants were led by a meeting facilitator through a number of computer-based activities such as to generate ideas, evaluate them in a variety of ways, do stakeholder analysis, and prioritize alternatives (Nunamaker et al., 1991, 1996/1997). But these systems fell into disuse because of their management overhead and cost.

Informal meeting support tools typically take the form of a simple projected interactive medium, such as a Word outline or a Google Doc. The outline lists the agenda items at the highest level in the outline; during the meeting, a scribe takes notes that everyone can view and implicitly vet. As agenda items are completed, the outline format allows the item to be collapsed, implicitly giving a visual sense of progress. Those applications that allow multiple people to author the shared document, such as Google Apps, are even more powerful in these settings. When there is a single scribe, that person typically is so busy that he or she is barred from contributing to the conversation. When there are multiple authors "live," while one scribe talks, others can take over seamlessly to enter notes on what they are saying. Additionally, these note-taking tools have been used very effectively in teams that include people for whom English is not their

---

[5] For more information on Assembla, see https://www.assembla.com/home [May 2015].

native language. The real-time visible note-taking is akin to "closed captioning" of the meeting.

Workflow and Resource Scheduling Routine tasks that require input or approval from a number of people benefit from a structured digital workflow system. A number of efficient online systems handle this type of flow. For example, a very successful workflow system supports the National Science Foundation grant submission, review, discussion, and decision-making process, notifying the appropriate players in the process at the appropriate time, giving them the tools and information they need, recording their actions, and sending the process on to the next in line. Although the rigidity of these systems can sometimes prevent their adoption, a number of such systems have succeeded (Grinter, 2000).

In some research endeavors, especially in the natural sciences where the expense of a large piece of equipment necessitates researchers sharing it, systems have been put in place to schedule time on the equipment. For example, time allocation for use of telescopes is managed with software systems created with the joint goals of being fair to those requesting time and maximizing the use of the equipment. Bidding mechanisms have been explored to optimize various aspects of the complicated allocation problem (Takeuchi et al., 2010). Various kinds of auctions have been tested to both create an equitable distribution of time and to prevent people from "gaming" the system (Chen and Sonmez, 2006).

Information Repositories

Whether a science team or larger group is co-located or distributed, it often needs to organize and manage shared information. The model of informally collaborating by sending people edited documents as attachments is common but fraught with challenges. Issues of version control and meshing of changes emerge. A better solution is to have a place where the single document resides as a shared file, with all the authors having access. Microsoft, for example, offers Sharepoint, an integrated set of tools selected for file sharing. It includes collections of websites and collaboration and information management tools (including tools for tagging documents for permissions and types and automatic content sorting). It also allows search through all the contents. To date, however, the system has not been widely adopted by research universities, which are using a range of different collaboration tools.

Another example of a system for shared editing and file management, but with a more fluid form, is Google Apps and Google Drive. The applications within Google Apps (documents, presentations, spreadsheets, forms, and drawings) each can be shared with others or placed into a folder, which also can be shared via Google Drive. This set of features gives the users flexibility, but without vetted "best practices," many are not using the applications effectively. The variety of different "cloud" technologies for document sharing is confusing to users (Voida, Olson, and Olson, 2013). As individual scientists and research institutions adopt various tools, the lack of interoperability sometimes forces scientists to revert to the "lowest common denominator" of sending documents as e-mail attachments (see Box 7-2).

Scientists who share data rather than documents face an additional set of challenges related to data quality, data-sharing, and database management (Borgman, 2015). If data are being collected by a science team or large group then the members have to agree, at the outset, what constitutes good quality data. Many large science groups have goals that include sharing data across sites. For example, in the early development of the Biomedi-

---

BOX 7-2
User-Centered Design for Collaboration Technologies

Technology intended to support virtual collaboration sometimes does not support it and even poses a barrier to collaboration (Crowston, 2013). Unless the technology is chosen or designed to both fit the users' needs and be easy to learn and use, it will not support the collaboration. A collaboration tool that requires extensive training, is difficult to use, does not fit collaborative activities, or does not work well with other technology is likely to interfere with collaboration and may eventually be abandoned. User-centered design can help technology adapt to the users, not vice versa.

Developing technology to fit the users' needs requires careful analysis of the users' tasks, infrastructure, culture, and overall work context. Beyer and Holtzblat (1998) outline the steps in such an analysis to ensure that the technology has the right functionality. They consider

1. communication flow,
2. order in which steps occur in the work,
3. artifacts produced and used in the work,
4. culture, including power and influence, and
5. physical layout.

Once these are made explicit, the people making the decision about what suite of technology to use (whether it be purchased or created) can brainstorm and then design the final solution.

When then designing the user interface to the various technologies in the suite, Norman (2013) proposes six principles:

cal Informatics Research Network (BIRN), the participants believed that progress on understanding schizophrenia would benefit from having a larger sample size of magnetic resonance imaging (MRI) images of patients, both with and without schizophrenia, doing various cognitive tasks while being imaged. A great deal of effort was spent in ensuring that the tasks that the patients performed were standardized and that the various imaging machines were calibrated. In other large groups of scientists, great care was given to developing a shared ontology of medical terms so that patient data could be aggregated from different locations and from different medical specialties, each of which had its own vocabularies (Olson et al., 2008).

In some domains of science, the laboratory notebook is a key tool for recording and vetting information. The researcher uses the notebook to keep a personal record of daily activities, such as tests run, information gathered, and observations. It is important to sign and date each entry

1. Consistency: Similar technologies should work in similar ways; users should not have to learn new procedures for each new piece of software.
2. Visibility: Controls should be clearly marked and not hidden from user view.
3. Affordance: Form and other visible attributes of the technology should intuitively guide function, (e.g., clickable elements of the interface should be highlighted).
4. Mapping: There should be a clear and evident relationship between controls and their effects (e.g., as when volume on a slider bar increases if the bar is moved up or to the right).
5. Feedback: Effects should follow actions immediately and obviously.
6. Constraints: User options should be restricted when unavailable or inappropriate (e.g., grayed out when not allowed).

Whittaker (2013) notes that successful use of technology often relies on following best practice, but it is unclear how users are expected to learn best practice. A single system seldom does everything a group or team needs: one is for workflow and scheduling, whereas another is for storing and sharing information. Interoperability problems abound, as when data-sharing tools to do similar work operate in different ways. Users are not equally familiar with the components that make up systems, and frustration can cause people to fall back to lowest-common-denominator technologies such as e-mail or spreadsheets.

Research is needed to improve the design of collaborative technology for team science. Such design would benefit from the philosophy outlined in the Human Systems Integration approach that puts the human at the center (National Research Council, 2007a).

to record important discoveries, often feeding into patent applications. Noting the value of being able to store and share these notebooks, some large scientific collaborations have developed electronic notebooks. The Electronic Laboratory Notebook (ELN) developed at the Pacific Northwest National Laboratory (PNNL) was so well designed that it was used heavily throughout the labs and adopted by other collaboratories even in different domains (Myers, 2008).

Aspects of the Computational Infrastructure

The System Architecture   Many large groups of scientists have no choice as to how to architect their systems. The large-scale computation technology is either local or hosted on a private grid of secure machines, and, at NSF-funded centers, the data, often large, are stored on their own massive servers. At a more fundamental level, only a few large research projects can afford to create their own data storing and sharing systems; many scientists still rely on Microsoft Excel software.

Those scientists who have no need for storing or computing with massive data have a choice of whether to purchase applications for installation on their machines or to opt for computing and storage "in the cloud." If choosing to work in the cloud, then connectivity is important if collaborative access in real-time is required. Many cloud-based applications offer some level of off-line activity, although the availability of up-to-date version control is lost. A more serious concern for some is security. There is resistance to cloud computing among clinicians, military contractors, police and fire departments, certain government agencies, and others who are sensitive to information loss.

One interesting consequence of these different architectures is that each architectural choice creates its own behavioral consequences. When the applications and documents are on private machines, the mode of collaboration is hand-off, serial revision: Documents are revised with "tracking changes" on and sent to the author-editor, who in turn can choose to accept each change or not. The power resides in whoever the collective has made editor. In contrast, where the document and application resides "in the cloud," there is an implied place where those designated as editors can go to make changes. In this model, each edit appears as if accepted; the document is changed. Others can view the revision history and undo the changes, but at least at present, a reversion to an earlier version undoes *all* changes, not just one at a time. Neither model in its current form is ideal.

These are two entirely different modes of collaborating in terms of workflow. Often collaborators tacitly make the decision about who has the power to make changes, who can merely comment, and who has the final say in accepting the changes proposed. The existence of these two models

presents additional challenges to the users who are involved in collaborations of both kinds. They have to remember where something is stored, how
to find it, and who has the power to decide on edits in each case, a situation
referred to as "thunder in the cloud" (Voida, Olson, and Olson, 2013).

The Network   Underlying all collaboration technologies is the network.
Simply put, the bandwidth has to be sufficient for the kind of work to be
done. Most of the developed world has adequate bandwidth for ordinary
tasks, including video. Specialized needs that require large amounts of
bandwidth will require specialized network infrastructure. Many large
scientific projects have had to build high-performance networks to handle
the volume of data that comes from their instruments as well as specialized computing to garner enough resources to do the computation on that
mass of data. For example, the ATLAS detector at CERN produces 23
petabytes[6] of raw data per second. This enormous data flow is reduced by
a series of software routines that lead to storing about 100 megabytes of
data per second, which yields about a petabyte of data each year. A special
infrastructure is required to manage data flows of this size.

Large-Scale Computational Resources   In many areas of endeavor, such
as advanced scientific research or data mining in business, large-scale computational resources are needed. Certain high-end centers, such as the
National Center for Atmospheric Research, have traditionally developed
their advanced computational resources in-house. But organizations such
as NSF, realizing that there is a need for advanced computing in many areas
they serve, have supported the building of infrastructures to support advanced computation. The historically important supercomputer centers are
one manifestation. A particularly noteworthy example of advanced infrastructure to support such needs is the Grid, a sophisticated computational
infrastructure that is widely used (Foster and Kesselman, 2004). A more
recent example is the NanoHub,[7] a special computational infrastructure for
nanoscience and nanotechnology.

Human Computation   There is also a tradition of using human capabilities
aggregated over large numbers to achieve important computational outcomes, often called "crowdsourcing." Although there are examples of this
as early as the 1700s, the phenomenon has experienced a recent renaissance
under other rubrics (Howe, 2008; Doan, Ramakrishnan, and Halevy, 2011),
such as collective intelligence (Malone, Laubacher, and Dellarocas, 2010),

---

[6]A petabyte is 1015 bytes. As reference, 103 = kilobyte, 106 = megabyte, 109 = gigabyte,
1012 = terabyte.

[7]For more information, see http://nanoHUB.org [May 2015].

the wisdom of crowds (Surowiecki, 2005), and citizen science (Bonney et al., 2009; Hand, 2010). The core idea is that in many domains, gathering together the small inputs of a large number of individuals ("micro tasks") can lead to results that can be as high in quality as judgments by experts and done in a fraction of the time.

In sum, science groups or teams typically need technologies to support **communication** and the sharing of the objects around which conversations take place. Technologies are needed to **coordinate** the conversations, both to find times to converse and to coordinate around the objects. The objects, **information**, and/or data, need to be collected to exacting standards, managed, and made accessible. Underlying it all is the **architecture** and **networking**, and large-scale **computation** occasionally supplemented by aggregated human computation. Effectiveness happens when the tools needed are available and used appropriately by the group or team members.

Selecting a Constellation of Technologies to Meet User Needs

New technologies often fail to live up to their promise, and it is not always clear what underlies the success of certain technologies, though these factors seem to include active leadership, deployment strategies, and how a particular tool fits in an overall assemblage of tools (Whittaker, 2013). Therefore, which technologies are chosen for a particular science team or group, and how these technologies are managed, can have an impact on the success of the collaboration. In selecting a constellation of technologies for a virtual team or group, it is important to consider the following factors (Olson and Olson, 2014):

- speed of response, impacting conversation and immediacy of data understanding;
- size of the message/data or how much computation is required, impacting required computation and networking;
- security, impacting choices about architecture;
- privacy, again, impacting choices about architecture;
- accessibility, impacting who can easily get access;
- richness of what is transmitted, impacting conversation and data understanding;
- ease of use, impacting adoption;
- context information, impacting coordination across sites;
- cost, impacting what can be accomplished; and
- compatibility with other things used, impacting adoption.

Choosing the appropriate suite of technologies to support a science team or group is not easy. The features of each technology drives how it will be used and often dictates social configurations of use. Although we have not provided a decision tree to guide selection of the "right" set of technologies, we have provided a listing of classes of collaboration technologies and the key features of these technologies that should be carefully considered in the choice of one's particular use. It is important to consider all facets of collaboration at a distance: communication, coordination, information repositories, and computational infrastructure.

HOW TECHNOLOGY AND SOCIAL PRACTICES CAN ADDRESS THE CHALLENGES OF VIRTUAL COLLABORATION

We next consider some examples of how technology and particular social practices can address each of the challenges we have identified to remote collaboration

Being Blind and Invisible

Videoconferencing and awareness tools can be used to increase visibility of participants as well as display who is working on what. Because it is important to communicate explicitly about the nature of work to be done as well as to share contextual information surrounding the work, videoconferencing can provide a feeling of presence for remote members and permit gestures, linguistic cues, and other ways to enhance communication among virtual team members. Awareness tools that permit the use of status indicators (such as IM) or color coding of document changes (such as Google Docs) can also be beneficial. Of course, they are only effective if the people involved invoke them, keeping the video on, setting their status markers to indicate their availability, and using the issue tracking systems.

Time Zone Differences

Whether a few hours or a full working day apart, scheduling meetings and coordinating work across time zones can be a challenge. A shared calendar, when used by all members of a virtual team, can greatly reduce the time spent scheduling teleconferences and work-related conversations. The calendar can signal when team members are working on parts of the task in addition to highlighting when they have free time available for casual conversations about the work. The shared calendar also serves as a form of documentation of the times members regularly meet, and especially for those across time zones, can reinforce norms around regular meetings.

Scheduling meetings including people whose workdays do not overlap can still create imbalance in who has to be inconvenienced. This is a social issue that has to be worked out with the participants and their management.

Differences Across Institutions

Typically, group members at different institutions are subject to different protocols, database access, and calendars. Workflow and resource scheduling that incorporates different institutional priorities, policies, and procedures can make coordination needs of participants salient. When group members from two different institutions do not share the same academic calendar, have different protocols for Institutional Review Board approval, or have different levels of access to online databases, coordination challenges can arise. Through a workflow and resource scheduling system that documents which group members are responsible for which tasks, who has access to particular sources of information, and what approvals are required and when, the institutional differences can be made explicit and accomodated. Systems that allow members and leaders to keep track of activities across institutions and provide notifications when action is required should facilitate coordination for multi-institution science groups.

Coordinating work around all of these differences requires explicit discussion. Successful distance collaborations often begin with a "communication covenant" that outlines the differences across institutions and the procedures the participants have agreed upon to coordinate.

Differences Across Countries

One of the best tools for determining how laws, rules, and policies vary between countries is a broadly accessible information repository such as a wiki. Groups that use such information repositories can document and track changes in regulation and intellectual property laws as they are occurring. Because all members have access to the latest information posted on the wiki, and can add, modify, or delete as necessary, the task of keeping national information up to date is shared across group members.

Differences Across Cultures

Today, English is the *lingua franca* of international scientific collaboration involving U.S. institutions. Much confusion and misunderstanding can follow from an understandable failure to appreciate linguistic nuances especially when spoken by remote members of large groups. Written communication, through e-mail, texting, and "chat rooms," allows people to write out what they are thinking, and, furthermore, allows other members to read (and re-read) the message to process what it means. Members from different cultures might find text-based communication more effective than real-time, voice-based communication.

In addition, suites of tools such as GlobeSmart have been designed to educate people about their collaborators' cultures and behaviors and to find a middle ground.[8] For example, if one is from a culture where the manager typically makes important decisions, she or he will be surprised when a collaborator hesitates in agreeing with the manager because everyone is consulted before a decision is made in the other culture.

Uneven Distribution of Members Across Participating Locations

Skillful use of meeting support technology can facilitate and broaden participation in decision making (e.g., by distributing a dynamic agenda), build procedural fairness (e.g., through electronic voting) across sites, and reduce power differences. When a majority of members are at the headquarters with a few other members scattered across different sites, it is easy for the remote member to feel isolated and in the minority. Meeting support technology, such as having a common Word document with an agenda that gets annotated as the meeting progresses, can ensure that members from all locations get heard (and recorded). A PowerPoint slide that outlines the procedures for voting on a decision, or even indicates who is going to lead the meeting (which can switch each time), can put the virtual group or team on the same page. The use of WebEx and other tools for running distributed meetings that integrate voice, documents, slides, and other materials facilitate the inclusion of members from different sites, big and small. These tools exist; it takes a manager open to contributions from all participants to use the tools effectively.

SUMMARY, CONCLUSIONS, AND RECOMMENDATIONS

Large groups of scientists, as well as smaller science teams, are often geographically dispersed, requiring scientists to rely on information technology and other cyber infrastructure to communicate with distant teammates. Addressing the special challenges facing such teams requires effective leadership and technology.

CONCLUSION. *Research on geographically dispersed teams and larger groups of scientists and other professionals has found that communicating progress, obstacles, and open issues and developing trust are more challenging relative to face-to-face teams and larger groups. These*

---

[8]For more information, see http://www.globesmart.com/about_globesmart.cfm [May 2015].

limitations of virtual collaboration may not be obvious to members and leaders of the team or group.

RECOMMENDATION 4: Leaders of geographically dispersed science teams and larger groups should provide activities shown by research to help all participants develop shared knowledge (e.g., a common vocabulary and work style). These activities should include team professional development opportunities that promote knowledge sharing (see Recommendation #2 earlier). Leaders should also consider the feasibility of assigning some tasks to semi-independent units at each location to reduce the burden of constant electronic communication.

CONCLUSION. *Technology for virtual collaboration often is designed without a true understanding of users' needs and limitations, and even when a suite of appropriate technologies is available, users often do not recognize and use its full capabilities. These related problems may thus impede such collaboration.*

RECOMMENDATION 5: When selecting technologies to support virtual science teams or larger groups, leaders should carefully evaluate the needs of the project, and the ability of the individual participants to embrace new technologies. Organizations should promote human-centered collaboration technologies, provide technical staff, and encourage use of the technologies by providing ongoing training and technology support.

8

Institutional and Organizational
Support for Team Science

This chapter addresses institutional and organizational support for team science. Following a brief preface, the first section introduces the organizational perspective. The second section focuses on the role of the research university in supporting team science. The third section discusses various organizational contexts for team science. The fourth section addresses how design of physical space may influence team science, and the chapter ends with conclusions and a recommendation.

Factors at the organizational and institutional[1] level influence the dynamics and effectiveness of science teams and larger groups, but research on these factors is limited. Recently, several scholars have highlighted the importance of these factors. For example, O'Rourke et al. (2014, p. 291) proposed that "the relationship between a collaborative, interdisciplinary research project and its context is a key determinant to project success." Stokols et al. (2008b) identified several organizational factors as important for motivating members of science teams—including strong incentives to support collaborative teamwork; non-hierarchical structures to facilitate team autonomy; and a climate of sharing information, credit, and leadership. Bennett and Gadlin (2014) drew on theories of social identity (how people think about themselves relative to a larger community) and procedural justice in organizations to argue that effective interdisciplinary

---

[1] Social scientists define "institutions" as enduring systems of established and prevalent social rules that structure social interactions (Hodgson, 2006). They define an "organization" as a type of institution that has established boundaries, a differentiated division of labor, and an integrated structure of coordination and control, for example, universities and business firms.

177

collaboration requires establishing trust between scientific teams and the organizations that house them. The authors viewed trust as the foundation for articulating an organizational vision, implementing change supportive of team science, and managing conflict.

However, few of these organizational factors have been scientifically studied to determine their relationship to the effectiveness of team science. It has been noted by several researchers (e.g., Luo et al., 2010) that empirical research into the institutional infrastructure of scientific research is rare. Winter and Berente (2012) argued that it is impossible to understand the goals of team science projects without considering how project goals are related to the goals of project members' home institutions, for example, academia, medicine, the law, capitalism, and engineering. Although these institutional goals influence project members' daily practices and their motivation to pursue the project goals, researchers have given "a dearth of attention to the contexts within which teams operate" (Winter and Berente, 2012, p. 443). Similarly, noting that the structures of research organizations have changed dramatically in recent years, Cummings and Kiesler (2011) called for applying organizational theory to these new arrangements, to enhance understanding of them, guide science policy, and refine theory.

THE ORGANIZATIONAL PERSPECTIVE

Conducting a full review of the large literature on organizations in terms of its relevance to team science was not possible within the time frame of the study. Here, we briefly review a few relevant studies, noting that they are predominately theoretical and case-study based, in contrast to the empirical and larger-scale studies of individual- and team-level factors reviewed in the previous chapters.

One facet of the ongoing debate in the organizational sciences about the relationship between organizational strategy and organizational structure (e.g., Hall and Saias, 1980; Mintzberg, 1990) considers how organizations can foster innovation through research and development. For example, in an early study, Burns and Stalker (1961) argued that "mechanistic" hierarchical organizational forms and management approaches were suitable for stable industries, while "organic" approaches with more fluid definition of functions and lateral interactions among peers were more suited to rapidly changing, research-intensive industries. Lawrence and Lorsch (1967) argued that successful organizations balance differentiation into functional departments (such as manufacturing, marketing, and research and development) with integration and collaboration across departments. Departments performing more stable tasks, such as manufacturing, had a more hierarchical structure than research and development departments performing rapidly changing tasks.

Focusing specifically on science, Shrum, Genuth, and Chompalov (2007) examined large, multi-institution groups of scientists in the fields of space science, oceanography, particle physics, and geophysics. The authors identified four types of organizational structures among these groups: bureaucratic, leaderless, non-specialized, and participatory. They proposed that the type of structure depended partly on the data collection methods and scope of research activities (i.e., the research strategy). For example, the highly participatory structures of particle physics resulted from the very large numbers of scientists who could collect data only by sharing access to a few particle accelerators, and a broad scope of collaborative activities. More generally, Shrum, Genuth, and Chompalov (2007) found that some degree of formal organization and management enhanced success across all four structures, including the non-hierarchical participatory ones. Surprisingly, given the longstanding scientific tradition of individual autonomy, participants in these large groups valued bureaucratic organizational structures that protected their rights to acquire and use data and prevented any one unit or institution from imposing its interests on the others. Such structures also handled purchases of large amounts of instrumentation, freeing scientists to focus on data collection and analysis. Large groups engaged in innovative technology or difficult logistical challenges benefited from employing professional project managers to deal with budgets and scheduling.

Another strand of organizational research relevant to team science has focused on management to foster innovation. For example, Simons (1995) argued that traditional, hierarchical management systems were obsolete and that, to foster innovation and effectiveness, managers should deploy four "levers of control":

1. *Belief systems* that employees internalize in response to ongoing leadership efforts to communicate core values through mission statements, credos, and vision statements.
2. *Boundary systems* that define the limits of freedom, such as codes of conduct and ethics statements.
3. *Diagnostic control systems* that are the traditional systems firms use to monitor and adjust operating performance, such as business plans, budgets, and financial and cost-accounting systems.
4. *Interactive control systems* that provide strategic feedback and guidance to update and redirect strategy such as competitive analysis and market feedback reports.

Similarly, O'Reilly and Tushman (2004) described how an "ambidextrous" management approach can help a company become adaptive and innovative, yet at the same time, efficient. Likewise, Adler and Chen (2011) argued that organizations engaged in large-scale creative collaboration need to help individuals balance the dual challenges of demonstrating creativity and embracing the formal controls that coordinate their creative activities with the activities of others. This suggests that organizations housing science teams (e.g., research centers, national laboratories, universities, private firms) would benefit from helping scientists to think creatively not only about their own, specific research projects, but also about how to best coordinate their efforts with others to advance organizational goals.

Based on an extensive review of the literature on management of research and development and other creative activities, along with motivation theory and identity theory, Adler and Chen (2012) suggested that two types of motivation are most important for creative tasks: *intrinsic motivation* and *identified motivation*. Intrinsic motivation refers to the voluntary willingness to engage in a task for the inherent pleasure and satisfaction derived from the task itself (Muyarama et al., 2010). Identified motivation reflects one's feelings of identity with a group or organization and motivates one to work toward collective goals. The authors proposed that organizations can foster these motivations by adopting human resource policies designed to attract and retain individuals with either high intrinsic motivation or fluid motivation (which is open to organizational influences), and by applying Simon's (1995) four levers, summarized above.

The authors proposed that organizations wishing to foster collaborative creativity also provide incentives combining individual and team rewards, as team rewards have been shown to encourage creativity (Teasley and Robinson, 2005; Toubia, 2006). They noted an experiment by Chen, Williamson, and Zhou (2012), which found that group-based rewards led to increased creative performance, as well as greater group cohesion and collaboration and increased identification with group objectives.

This brief review of theory and research has potential implication for science teams and for the organizations that house them. The studies reviewed have explored how to manage task uncertainty in rapidly changing environments, which is characteristic of scientific work, particularly in the early stages of developing a research project. Similarly, the various authors highlighted the need to manage interdependence, which is characteristic of science teams, especially interdisciplinary and transdisciplinary teams (Fiore, 2008). However, much further research is needed to more clearly articulate the connections between organizational theory and research and team science.

UNIVERSITY POLICIES AND PRACTICES

Experts in higher education studies view universities as complex organizations composed of multiple, loosely coupled subsystems (Austin, 2011). Faculty members work within various contexts and cultures—including the department, the college, the institution as a whole, and external groups, such as disciplinary societies and accrediting associations—that can be conceptualized as "levels" of the university organization. These various contexts and cultures influence faculty attitudes and choices about research, teaching, and service, including their attitudes and decisions related to team science. Within these complex systems, some of the key factors influencing faculty behavior include evaluation and reward systems, workload allocation, professional development opportunities, and leadership. Multiple factors at multiple levels of the system simultaneously influence faculty member choices and behaviors. Given that higher education institutions are complex organizations, change efforts are most effective when they use both a "top-down" and a "bottom-up" approach, take into consideration the factors at work within the multiple contexts that affect faculty work, and strategically utilize multiple change factors (Austin, 2011). With this perspective in mind, we now turn to a discussion of how universities are working to support team science.

University Efforts to Promote Interdisciplinary Team Science

Many experts view current university policies and discipline-based organizational structures as an impediment to interdisciplinary team science. For example, Klein et al. (2013, p. 1) argued that "obstacles to . . . [interdisciplinary team science] span the entire academic system of organizational structure and administration, procedures and policies, resources and infrastructure and recognition, reward, and incentives." In an earlier study, Klein (2010) called for a comprehensive, university-wide approach to remove obstacles to interdisciplinary research and teaching among faculty who are part of the entrenched disciplinary culture and organization of research universities.

In contrast to these views, universities around the country have recently launched many efforts to promote interdisciplinary team science (see Duderstadt, 2000; Frodeman et al., 2010; Klein, 2010; Altbach, Gumport, and Berdahl, 2011; Repko, 2011; O'Rourke et al., 2014; among others). University leaders have created new science teams, larger groups, and research centers, encountering the benefits and challenges of diverse membership and deep knowledge integration, while also generating new challenges of goal alignment among the new teams and other entities. One example, among many, is Arizona State University (ASU). In the past decade, under the leadership of President Michael Crow, ASU has become a national pacesetter in restructuring the university to promote interdisciplinary team research and teaching (Crow and Debars, 2013; Martinez, 2013; see also http://newamericanuniversity.asu.edu [May 2015]). Using a top-down, institutional redesign approach, the university has built new interdisciplinary schools and research centers, including a School of Biodesign, a School of Sustainability, a School of Human Evolution and Social Change, and a Beyond Center. These efforts have attracted much research funding, many students, and highly qualified faculty to the university, but sometimes with the costs associated with frequent organizational restructuring of academic units.

The University of Southern California (USC) has adopted a more bottom-up approach to supporting team science, creating a fund to provide seed grants to interdisciplinary projects selected by a faculty committee and revising its promotion and tenure policies with faculty involvement, as discussed further below. It will be interesting to see how these different approaches at USC and ASU play out in a longer time perspective, and if one is more effective than the other in promoting academic culture change over time. It also will be important to see how these changes not only directly affect team science research, but also student training, because, as Austin (2011) cogently argued, "doctoral socialization" by Ph.D. advisors in the training of prospective faculty members strongly influences how the next generation of faculty view teaching and research, including team science. M. Duane Nellis, president of the University of Idaho (2013, p. 226), calls for both approaches, arguing that efforts to promote transdisciplinary research "must be led both from administrators at the top and from a broad spectrum of faculty at the base." However, he also cautions that implementation of administrative policies and procedures is uneven, due to the influence of traditional departmental and disciplinary boundaries and cultures, and the lack of funding for cross-departmental research efforts (e.g., in the form of research assistantships).

Many other examples of efforts to promote interdisciplinary team science can be found at campuses across the United States. Northwestern University, under the leadership of former President Henry Bienen and continuing to today, provides a good example. Bienen fostered ties to the Argonne National Laboratory and to the Chicago biomedical community, as well as stimulating and supporting interdisciplinary team science on campus. In another example, Rutgers University President Robert Barchi is encouraging interdisciplinary research by placing several "catalysts" throughout the university, including creating a new position, director of research development, within the Office of the Vice President for Research (Murphy, 2013). Barchi has also merged two medical schools, a nursing school, and a school of applied health professions onto the main Rutgers campus, fostering an intermingling of faculty that has led to growing interdisciplinary team science efforts.

Promotion and Tenure Decisions and Team Science

Although scientists are motivated by a variety of factors, including prestige and the freedom to pursue their individual research interests (Furman and Gaule, 2013), one important factor is money. Thus, an important way universities can support team science is by recognizing and rewarding individuals for their team-based accomplishments when granting tenure. Decisions about promotion and tenure are typically made by faculty committees within disciplinary departments, with review and approval by the dean of the relevant school and higher-level administrators. These decisions are affected by current trends and more enduring scientific norms.

One important trend is the decline (in real terms) of total federal and state funding for scientific research (National Research Council, 2012a). In biomedicine, for example, based on the expectation that past funding increases for biomedical research would continue indefinitely, universities have created more and more research positions that depend on temporary grants (often referred to as "soft money"). They have continued in an ever-more intense competition for a shrinking pool of federal dollars (which do not cover all costs of research) while also responding to federal and state regulatory and reporting requests that impose burdensome monetary and time costs (National Research Council, 2012a; Alberts et al., 2014). These financial problems discourage universities from providing tenure.

Another, partially related trend is the decline of tenure. The percentage of degree-granting institutions with tenure has declined from 63 percent in the 1993–1994 academic year to 45 percent in the 2011–2012 academic year (U.S. Department of Education, 2013). In 1969, 78 percent of faculty members were tenured or in tenure-track positions; by 2009, only 34 percent of faculty members were in tenured or tenure-track positions (Kezar and Maxey, 2013). Tenure rates even within the ranks of only full-time instructors have also declined—from 56 percent in the 1993–1994 academic year to 49 percent in the 2011–2012 academic year (U.S. Department of Education, 2013). Replacing tenured and tenure-track positions are "adjunct" positions, staffed by instructors who may be hired on 1-year contracts or paid by the course (Kezar and Maxey, 2013).

While these two trends reduce tenure prospects for all young scientists, enduring scientific norms may pose special obstacles to candidates seeking tenure for team science.

In his classic studies of the "Matthew Effect," Merton (1968, 1988) found that more eminent coauthors tended to receive disproportionately more credit for team-authored work than their less eminent coauthors. The Mathew Effect can also work in reverse. Jin et al. (2014) investigated how retractions (papers recalled because of errors) affect trust in an author's prior work as measured by citations to the author's prior publications. They found that scientific misconduct imposes little citation penalty on eminent coauthors, but less eminent coauthors face substantial citation declines to their prior work.

The Matthew Effect suggests that in assessing authors' contributions to a collaborative paper, the scientific community presumes that the more eminent coauthor deserves the lion's share of the credit, whereas the other co-authors are relegated to subordinate roles. Merton noted that this pervasive credit assignment mechanism is likely to affect scientists' career advancement and motivation for working in teams.

Faculty members charged with making tenure decisions are influenced by these current trends and norms. Pressed for time because of the competing demands of service on the tenure committee and their own research and teaching, they may not thoughtfully read the candidate's scholarly publications, but rather seek shortcuts, in the form of simple metrics to assess the quality and importance of the candidate's work (Tscharntke et al., 2007). For example, they may focus primarily on whether the candidate has published in the most prestigious journals within the relevant field or on the "impact" of the candidate's publications (the number of times the publication is cited by others). When asked to evaluate a candidate's contributions to team research, as reflected in multi-authored publications, committee members face additional challenges, including potential bias resulting from the Matthew Effect (Merton, 1968). Disciplinary norms for assigning credit based on the order of the authors' names may not help in assigning credit for interdisciplinary publications. In addition, Tscharntke et al. (2007) noted that, beyond the widely accepted norm that the first author should receive most credit, norms for assigning credit in multi-authored publications vary widely across research fields and countries.

Current Status of Promotion and Tenure for Team Science

Systematic data about the extent to which candidates do or do not win tenure on the basis of team science research are lacking. However, respondents to surveys conducted as part of an earlier National Academies study (National Academy of Sciences, National Academy of Engineering, and Institute of Medicine, 2005) ranked promotion and tenure criteria the highest of the five impediments to interdisciplinary research. Based on a literature review on promotion and tenure policies and practices affecting interdisciplinary team science (Klein et al., 2013), Professor Julie T. Klein, Wayne State University, told the committee:

> The current picture across campuses, however, is more mixed. Risks differ by field and by institution. Furthermore, a growing body of precedents, guidelines, and models are available. Individuals are still too often vulner-

able, however. An old saw continues to haunt prospects for tenure and promotion: "Tenure first, interdisciplinarity later. . . . Its counterpart in team science is 'Individual reputation first, collaboration later.'"

Echoing similar concerns, the United Kingdom Academy of Medical Sciences has launched a study of incentives and disincentives for participating in team science (Academy of Medical Sciences, 2013). Taken together, these various reports indicate that uneven evaluation of tenure candidates' contributions to team science projects poses a barrier to their chances of winning tenure.

University Policies for Supporting Team Science Through Tenure and Promotion

No systematic, national data are available on university policies designed to help promotion and tenure committees recognize and reward team science. However, a recent survey by Hall et al. (2013) provides some suggestive evidence. The survey asked 60 institutions receiving Clinical and Translational Science Awards from NIH about their tenure and promotion policies. The authors noted that this is a biased sample, because the center awards are specifically designed to support translational team science and grantee institutions are therefore more likely than other institutions to recognize team science in their policies. Of the 42 institutions that responded, 10 indicated that their promotion and tenure guidelines did not include any language specific to collaborative, interdisciplinary research and/or team science, while 32 did have such language. Among the 32 guidelines with such language, most included small modifications to traditional promotion and tenure criteria and primarily focused on issues of authorship (e.g., suggestions to annotate the candidate's bibliography to substantiate middle-authorship roles). Only a handful offered alternative criteria meant to capture contributions unique to the team science. These criteria were vague and did not include indicators or metrics of attainment, relying instead on written statements by the candidates and their collaborators. The authors called for further research and development of actionable criteria to assess individual contributions to team science. In particular, they called for research to better understand contributions made by scientists that advance scientific research through actions and roles other than authorship.

As indicated by the survey, some universities are providing more guidance to departments, deans, and tenure and promotion committees than in the past for evaluating scientists involved in interdisciplinary and team science research. In doing so, they face the challenge of not only developing high-level goals or policy statements, but also implementing or aligning these goals with the culture of departments and individual faculty members at lower levels within the university system. The following example illustrates how USC built a new approach from the bottom up.

The USC (2011) guidelines for assigning authorship and attributing research contributions provide straightforward principles and policies for evaluating individual scholarly contributions to research and publication. Developed by faculty committees following a series of six workshops on collaboration and creativity (see Berrett, 2011) and approved by the university's academic senate, the guidelines deserve to be the starting point for discussions at campuses around the country. The guidelines (University of Southern California, 2011) commit USC to four strong principles:

1. fair and honest attribution of the contributions of each person in the creation of research products and creative works;
2. allowance for diversity in the attribution of contributions, which vary across disciplines and dissemination outlets;
3. making our research products and creative works readily available to others, so that they may be further developed or implemented; and
4. avoidance of disputes over attribution and ownership that may create impediments to the creation and dissemination of significant and impactful research, scholarship, and creative works.

The guidelines further clarify the types of contributions required to qualify as an author and ask team members to decide among themselves about the order of author names, acknowledging that conventions for order of authorship vary across disciplines.

New policies such as those at USC are unusual, and most of the available evidence indicates that university policies typically lack clear criteria for evaluating an individual candidate's contributions to team-based research. To address this problem, the committee recommends at the end of this chapter that universities and disciplinary associations develop broad principles and more specific criteria for tenure committees' use when allocating individual credit for team-based work, echoing the recommendation of a recent National Research Council (NRC) report on transdisciplinary research, or "convergence" (2014).

Recent Developments in Authorship Attribution

In a recent development that could assist universities in the difficult challenge of allocating credit for team-based work, major journals, such as the *Proceedings of the National Academy of Sciences, Nature,* and the journals published by the Public Library of Science, have begun to require an "author contributions" section describing each author's contribution to the published article. Such sections represent a potential step forward from relying on varying authorship conventions to determine how much credit each author deserves for a publication. Tscharntke et al. (2007) proposed that, when preparing these "author contributions" sections, the authors should explicitly identify the authorship convention to be used in allocating credit for the work, such as stating that the authors are listed in order of importance of contribution or that all authors contributed equally. To simplify and standardize the process of describing all contributions, Allen et al. (2014) developed a preliminary taxonomy of 14 contributor roles, ranging from study conception to providing resources. Such a taxonomy could be included in manuscript-submission software, allowing researchers to easily assign roles in the process of writing and submitting the paper. Two of the authors of the Allen et al. (2014) taxonomy have launched a project to further develop, maintain, and implement it, in collaboration with publishers, funding agencies, researchers, and university administrators (CRediT, 2015).

Another new approach would build on the emerging databases of scientific authors and publications, such as the research networking systems discussed in Chapter 4. Such databases allow scientists to interact, form networks and interest groups, and rate each other's publications. New software additions to these systems could allow multiple authors of a paper to publish descriptions of each member's contribution, and each contributor could verify what others contributed (Frische, 2012). If widely accepted, then these types of systems would be helpful to scientific journals, funding agencies, and university promotion and tenure committees.

Individual and Team Rewards

Awarding tenure is only one component within the larger academic and scientific system of rewards and incentives. The questions surrounding how to recognize individual contributions to team-based research in tenure decisions raise related questions about the possibility of recognizing and rewarding teams. As discussed earlier in this chapter, recent research suggests that team-based rewards support team creativity. In addition, Horstman and Chen (2012) have recently studied group-based rewards for individual and group contributions to solving scientific problems. Further research is needed on this topic.

ORGANIZATIONAL CONTEXTS FOR TEAM SCIENCE

Team science is conducted in a variety of organizational contexts that may be located within, outside, or span the boundaries of the research university. For example, government-university-industry partnerships may be organized as networks, research centers, or free-standing institutes. Here, we briefly discuss some of these contexts.

Research Centers

Over the past two decades, universities, businesses, and public and private funders have increasingly established research centers and institutes to support multiple, interrelated research projects focusing on a common theme.[2] In 2006 (the most recent year for which data are available), there were an estimated 14,000 nonprofit research centers in the United States (D. Gray, 2008). Centers and institutes often house interdisciplinary or transdisciplinary research and university-industry research partnerships. For example, a recent NRC study (2014) focused on "convergence institutes," which integrate life sciences, physical sciences, and engineering and forge industry partnerships to support the research and facilitate its translation. The study profiled institutes such as Bio-X at Stanford University, the David H. Koch Institute for Integrative Cancer Research at the Massachusetts Institute of Technology, and others. These and other transdisciplinary research centers encounter both the benefits and challenges of diverse membership, deep disciplinary integration, and large size.

Although only limited research is available on the processes and outcomes of research centers and institutes (Bozeman, Fay, and Slade, 2012), evaluations of federally funded centers provide some insights. For example, the National Science Foundation (NSF) launched the Science and Technology Centers (STC) Integrative Partnerships Program in 1987, in response to a call from President Reagan. Solicitations for center awards set the range at $1.5–4 million per year, for a maximum of 10 years. A recent review of this program by the American Association for the Advancement of Science (AAAS; Chubin et al., 2009) found that it was "an effective and distinctive mode of Foundation support for addressing grand challenges and emerging opportunities in science and technology" (p. 79). Based on analysis of multiple measures, including publication counts and participant surveys, the authors concluded that the STC program had succeeded in (1) connecting national priorities in science and engineering with "frontier" academic science and engineering research; (2) encouraging established researchers to venture into more risky areas; (3) bringing together different disciplines; and (4) fostering collaboration between basic and applied scientists. The authors also found that the program positively affected doctoral student training and the centers actively carried out "knowledge transfer" activities, ranging from publishing articles and creating new journals to supporting regional economic development through technological innovation.

---

[2]Chapter 9 provides data on the growth in NSF and NIH funding of research centers.

The review also noted weaknesses of program management. At the time of the review, the STC program did not belong to any single research directorate or office within NSF and was forced to compete for resources not only with the traditional individual-investigator mode of support, but also with directorate-based center programs. The matrix model of the organization was found to impede accountability, and the annual review process—a key tool used by NSF to monitor performance—was "vulnerable to changing, inconsistent and at times idiosyncratic advice from review teams" (Chubin et al., 2009, p. 84). Finally, reflecting the need for this study and the science of team science, the review found that the existing system for collecting and analyzing performance data was poorly suited to evidence-based decision making.

In 2006, NIH launched the Clinical and Translational Science Awards (CTSA) Program to "advance the assembly of institutional academic 'homes' that can provide integrated intellectual and physical resources for the conduct of original clinical and translational science" (Zerhouni, 2005, p. 1622).

The program built on the NIH General Clinical Research Centers Program, which had provided clinical research infrastructure funding, as well as funding programs for disease-specific centers. Under it, individual research centers are funded through 5-year cooperative agreements, with site budgets ranging from $4 million to $23 million annually. The Institute of Medicine (2013) found that the program has demonstrated progress in three crosscutting domains that are important to advancing clinical and translational science: training and education, community engagement, and child health research. The IOM committee recommended that the program continue to provide training, mentoring, and education as essential core elements, emphasizing innovative models that include a focus on team science. They also recommended that the program disseminate high-quality online offerings for essential core courses for use in CTSA centers and other institutions. If these recommendations are implemented, then such courses would help to provide the professional development for team science recommended in Chapter 5.

To address the promotion and tenure challenges discussed earlier in this chapter, IOM recommended that CTSA "champion the reshaping of career development pathways for researchers involved in the conduct of clinical and translational science; and ensure flexible and personalized training experiences that offer optional advanced degrees" (p. 116).

Like the AAAS review of the STC Program, the IOM review of the CTSA Program identified management weaknesses. Specifically, the authors found that program leadership has relied primarily on the efforts of individual centers (awardees) and their principal investigators, leading to a largely ad hoc structure and process for identifying next steps and overall management. They also found that NIH had provided direction primarily through the funding announcements, which had emphasized different key functions or priorities in different grant cycles. To address this problem, the report recommended that the National Center for Advancing Translational Sciences strengthen its leadership of the program through several steps, including conducting a strategic planning process, forming partnerships with NIH institutes and centers, evaluating the program as a whole, and distilling and widely disseminating best practices and lessons learned.

To more clearly determine the outcomes of its investment in large, transdisciplinary research centers, the National Cancer Institute has supported an ongoing program of research on the effectiveness of team science (e.g., Stokols et al., 2008a). The insights emerging from this research program are discussed throughout this report.

University-Industry Research Partnerships

Earlier sections of this chapter discussed the challenges faced by universities in developing, maintaining, and assessing the success of science teams and larger groups. In university-industry research partnerships, new problems emerge, including proprietary concerns and profit motives in the development of commercial products. Because of the complexity of partnerships between universities and businesses with different motives and organizational structures, Bozeman and Boardman (2013)[3] refer to them in a paper commissioned by the committee as "boundary-spanning research collaborations."

Bozeman and Boardman (2013) conducted an extensive review of the literature on university-industry research partnerships and industry-industry interdisciplinary research partnerships, building on the review by Bozeman, Fay, and Slade (2012) on similar topics. Both types of partnerships are often housed in research centers or institutes.

Bozeman and Boardman (2013) found that the inclusion of multiple disciplines in university-industry research collaborations increased productivity but also was associated with increased diversity of incentives and motivations. Perhaps to address these diverse motivations, partnerships including multiple disciplines were more hierarchical and formally structured than partnerships involving only a single discipline. More generally, the authors found that prior acquaintance and trust were key factors for success in university-industry research partnerships, and, where these elements were absent, creating formal structures and authorities helped to manage conflict

---

[3] After submitting this paper to NRC, the authors subsequently published a paper addressing many of the same issues, titled *Research Collaboration and Team Science, A State-of-the-Art Review and Agenda*; see http://www.springer.com/series/11653 [May 2015].

and improve effectiveness. However, they also discussed a study focusing on Australian university-industry cooperative research centers that found that the formal legal contracts establishing the centers were rarely enforced (Garett-Jones, Turpin, and Diment, 2010). Instead, researchers and organizations within the centers relied on informal social mechanisms, such as trust and reciprocity, to coordinate work. In the absence of legal sanctions, researchers who perceived breaches of trust became less enthusiastic about the collaborative work and some withdrew from the centers. This study suggests that it is important to enforce the formal structures and authorities created when establishing university-industry research partnerships.

Bozeman and Boardman (2013) identified three major gaps in the research on university-industry partnerships. First, research on effective management of such partnerships is underdeveloped, often identifying best practices that are local, and may not work robustly across different contexts and situations. The scant available literature suggests that managerial practices are "poorly thought out and haphazard" (Bozeman and Boardman, 2013, p. 65). Second, little research focuses on the "dark" side of boundary-spanning research collaborations. Research on the failures of these collaborations is scarce. Failure was most prevalent when *both* formal and informal management structures were weak or one or the other was absent. Third, although some research suggests that intellectual property disputes are a real source of failures in university-industry research partnerships, there is little empirical research that directly addresses this issue. The limited research available suggests that careful contract monitoring can help to address intellectual property disputes, but such monitoring is sometimes lacking (e.g., Garett-Jones, Turpin, and Diment, 2010).

Bozeman and Boardman (2013) concluded that much remains unknown about university-industry research partnerships. They argued that evaluating the performance of these large groups of scientists is difficult because of measurement challenges (as discussed in Chapter 2), but more importantly to the lack of any baseline comparisons. The authors note that it remains unknown whether the scientists collaborating within a particular partnership or center would be more or less productive working individually or with collaborators other than those involved in the partnership. As noted in Chapter 1, a study by Hall et al. (2012b) begins to address this challenge, using quasi-experimental methods to compare the research productivity of scientists participating in large research centers with that of scientists investigating the same topics, but working individually or in small groups unaffiliated with the centers.

Bozeman and Boardman (2013) suggested that more research is needed on (1) how scientists, universities, and firms choose research partners; (2) the reasons for failure in university-industry partnerships; (3) the role of partnership participation in developing the human capital of individual scientists (i.e., their knowledge and social networks); and (4) effective management strategies for these partnerships. To address these and other gaps in the research, the authors called for moving beyond descriptive and taxonomic case studies to more systematic field and quasi-experimental research designs and moving beyond individual impact studies (e.g., individual productivity) to a greater concern with institutional outcomes.

Clearly, further research is needed to improve the management of university-industry research partnerships, as well as centers and institutes that are primarily academic. One study (D. Gray, 2008) pointed to improvement-oriented evaluation approaches as a way to both understand and improve center management. The NSF Industry/University Cooperative Research Program has adopted an improvement-oriented approach that meets the needs of an important internal stakeholder—the center director. The new approach has placed an on-site evaluator at each center. The evaluator (usually a social scientist) is uniquely positioned as both a center participant and an evaluator to identify and share with the director emerging challenges and problems. In addition to serving as consultants to the directors and conducting ongoing surveys, the on-site evaluators have contributed to a volume of best practices that is available to the center directors and the public on the NSF website (Gray and Walters, 1998). The use of ongoing, improvement-oriented evaluation to enhance performance at the center or institute level is somewhat similar to team development approaches at the team level discussed in Chapter 3. For example, the Productivity Measurement and Enhancement System (ProMES; Pritchard et al., 1988) intervention, which measures performance and provides structured feedback, has been shown to improve team self-regulation and performance (Pritchard et al., 2008).

Universities can support university-industry research partnerships and other types of research centers by providing the leaders with formal leadership training, as recommended in Chapter 6. They can also encourage leaders and participants in newly formed research centers or institutes to articulate their expectations through written charters or collaborative agreements (Bennett, Gadlin, and Levine-Finley, 2010; Asencio et al., 2012). Such documents outline how tasks will be accomplished, how communication will take place, and how issues such as finances, data sharing, and credit for publications and patents will be handled.

Inter-Firm Research Partnerships

Research collaborations involving multiple companies may take various forms, including research parks, research and development alliances with formal contracts, and joint ventures. In their literature review, Bozeman and Boardman (2013) found that inter-firm research partnerships shared many of the challenges of university-industry research centers. For example, in interdisciplinary and transdisciplinary inter-firm research partnerships including multiple firms, a lack of formal authorities and structures was associated with failures and, although careful contract monitoring and enforcement were vital to success, they were not always present. In addition, the authors identified gaps in the literature on inter-firm research partnerships similar to those in the literature on university-industry research partnerships.

Research Networks

Formal and informal research networks play an important role in catalyzing and supporting team science. For example, informal networks of scientists are often based on prior acquaintance, which, as noted above, facilitates rapid development of trust and thus supports the effectiveness of science teams and larger groups. Cummings and Kiesler (2008) found that virtual collaboration among groups of scientists was more likely to be maintained when the scientists collaborated with colleagues they had worked with previously. Disciplinary and interdisciplinary scientific societies provide opportunities for scientists to develop networks of colleagues with similar interests, through conferences, meetings, and online discussion boards, but fewer opportunities are available for scientists to establish professional relationships across disciplines.

Research funders have catalyzed the formation of networks to develop research on interdisciplinary topics, such as the Network on BioBehavioral Pathways in Cancer (National Cancer Institute, 2015). In another example, the MacArthur Foundation used a network approach to foster interdisciplinary research on mental health and positive psychology. Kahn (1993) described the evolution of the network, including the development of close interpersonal and intellectual relationships among the geographically dispersed participants. He reported promising early results, including the development of new data banks and resources available to investigators everywhere, along with validated assessment instruments. One indicator of the promise of this approach was the foundation's subsequent decision to fund research networks focusing on other topics, including the transition to adulthood.

OPTIMIZING PHYSICAL ENVIRONMENTS FOR TEAM SCIENCE

Regardless of where collaborative scientific research is conducted, it requires supportive physical environments. According to Stokols (2013), the features of team environments can enhance or hinder team members' capacity to focus their attention on developing shared knowledge, effective communication, and positive affect.

Yet while it appears to be intuitively obvious that physical environments influence the nature of team science, Owen-Smith's (2013) review of the relevant research found surprisingly little empirical evidence to back up such an impression.

Among the studies that do address this issue, Stokols et al. (2008b, p. S100) noted that a study of interdisciplinary treatment teams in hospitals by Vinokur-Kaplan (1995) found that "members' ratings of physical environmental conditions at work, such as the availability of quiet and comfortable places for team meetings. . . were positively related to reported levels of interdisciplinary collaboration." Studies by Kabo et al. (2013a, 2013b) have shown that within buildings (and on particular floors), walking path overlaps among scientists also promote collaboration. There are also numerous studies of corporate workspace design (see, e.g., Steele, 1986; Brill, Weidemann, and BOSTI Associates, 2001; Becker, 2004; and Doorley and Witthoff, 2012, among many others) that relate productivity to architectural design. However, Owen-Smith (2013) argued that many other contextual factors beyond the physical environment, such as organizational reward systems (e.g., promotion and tenure policies), also influence scientists' motivation to participate in team science and therefore more systematic research is needed before firm conclusions can be drawn.

Anecdotally, it would appear that physical spaces that encourage interaction among scientists, from regular interchanges to chance encounters, help stimulate collaborative thinking and work. The Santa Fe Institute, for example, provides open spaces with plenty of comfortable chairs, sofas, and white boards; offices with glass windows facing open spaces; offices shared with scholars from different disciplines; abundant glass walls with available markers to encourage scientists to write algorithms they are discussing on the glass and not wait to return to their offices; and lunches and teas shared by everyone in common spaces. Directors of other research centers share similar impressions. For example, at the NRC workshop on Key Challenges in the Implementation of Convergence, Carla Schatz, director of the transdisciplinary BioX Institute at Stanford University, emphasized the value of creating a physical home for core faculty, with a good cafeteria and high-quality coffee. The building, she said, serves as both a gathering point and a recruiting tool for attracting scientists across disciplinary boundaries to join the Institute and advance human health.[4]

However, the relationship of these physical design factors with successful team science remains impressionistic and unconfirmed by rigorous study.

---

[4] See https://www.youtube.com/watch?v=JysIA-4fcA4 [May 2015]; National Research Council (2014).

Two recent studies that used experimental designs point toward the type of research needed on this topic. First, Catalini (2012) exploited the fact that multiple academic departments at the University of Pierre and Marie Curie (UPMC) in Paris were relocated over a 5-year period because of an asbestos removal project to examine the role of location on collaboration patterns in a precise way that enabled him to identify the casual influence of location on research collaboration. He found that random relocations that resulted in co-location encouraged collaborations and also breakthrough ideas across academic fields. Boudreau et al. (2012) undertook a similarly creative effort to understand the role of location in collaboration by conducting a field experiment in which they randomized researcher locations, finding that those in even briefly co-located environments were more likely to collaborate.

The research to date, which has primarily examined correlational relationships, suggests several findings: spatial design that emphasizes functional zones where scientists' walking paths consistently overlap (Kabo et al., 2013a) leads to increased interaction; increased interaction can lead to stronger collaborations; and such collaborations can help lead to scientific successes. There are growing data to support these general correlations (see recent studies by Toker and Gray, 2008, Rashid, Wineman, and Zimring, 2009, and Sailer and McCulloh, 2012, all cited by Owen-Smith, 2013), but translating these correlations to proven causal relationships generally remains to be achieved. In particular, further research is needed that considers the role of physical space as one factor among many that influence the extent and quality of team science.

SUMMARY, CONCLUSIONS, AND RECOMMENDATION

Science teams and larger research centers and institutes are often housed within universities. In these complex organizations, faculty members' decisions about whether and when to participate in team science are influenced by various contexts and cultures, including the department, the college, the institution as a whole, and external groups, such as disciplinary societies. Formal rewards and incentive structures, reflecting these various cultures, currently tend to focus on individual research contributions. Some universities have recently sought to promote interdisciplinary team science by, for example, merging disciplinary departments to create interdisciplinary research centers or schools, providing seed grants, and forging partnerships with industry. However, little is known about the impact of these efforts, while the lack of recognition and rewards for team science can deter faculty members from pursuing it.

CONCLUSION. *Various research universities have undertaken new efforts to promote interdisciplinary team science, such as merging disciplinary departments to create interdisciplinary research centers or schools. However, the impact of these initiatives on the amount and quality of team science research remains to be systematically evaluated.*

CONCLUSION. *University policies for promotion and tenure review typically do not provide comprehensive, clearly articulated criteria for evaluating individual contributions to team-based research. The extent to which researchers are rewarded for team-based research varies widely across and within universities. Where team-based research is not rewarded, young faculty may be discouraged from joining those projects.*

In a few isolated cases, universities have developed new policies for attributing individual contributions to team science. At the same time, research has begun to characterize the various types of individual contributions and develop software systems that would identify each individual's role during the process of submitting and publishing an article. This work can inform new efforts by universities and disciplinary associations.

RECOMMENDATION 6: **Universities and disciplinary associations should proactively develop and evaluate broad principles and more specific criteria for allocating credit for team-based work to assist tenure and promotion committees in reviewing candidates.**

This chapter illuminates the limited evidence about team science from an organizational perspective. For example, at a time of many university efforts to promote interdisciplinary and transdisciplinary team science, Jacobs (2014) argued that there are dangers attached to a wholesale move away from traditional disciplines. He suggested that the growing volume of research makes specialization inevitable, and he viewed disciplines as broad and dynamic, in contrast to interdisciplinary research, which may be narrow and specialized. Finally, he argued that research universities based upon interdisciplinary principles may be more centralized, less creative, and more balkanized than current, very successful research universities. Such views highlight the need for more research on the outcomes and impacts of current university efforts to promote team science.

Further research is needed to more clearly understand how alternative organizational structures, management approaches, and funding strategies influence the processes and outcomes of research centers and other large groups of scientists. In addition, further research is needed that moves beyond correlations to consider how the physical environment interacts with other environmental factors (e.g., reward structures, time pressures) to motivate and/or discourage collaborative team science.

9

Funding and Evaluation of Team Science

Organizations that fund and evaluate team science face a unique set of challenges that are related to the opportunities and complexities presented by the seven features that create challenges for team science first introduced in Chapter 1. Funding science teams and larger groups is different from funding individuals, and the differences increase when teams and groups include features such as large size, the deep knowledge integration of interdisciplinary or transdisciplinary projects, or geographic dispersion. Evaluating all phases of such complex teams and groups, from proposals to how the funded teams or groups are progressing to the project outcomes can be challenging. It requires an understanding of how teams or groups conduct science that leaders and staff members of science funding organizations may lack. Recognizing this problem, the National Science Foundation (NSF) commissioned the current study to enhance its own understanding of how best to fund, evaluate, and manage team science, as well as to inform the broader scientific community (Marzullo, 2013). The National Cancer Institute supports the new field of the science of team science for similar reasons, including to clarify the outcomes of its investments in large science groups (e.g., research centers) and to increase understanding within the scientific community of how best to support and manage team science (Croyle, 2008, 2012). In addition, a federal Trans-agency Subcommittee on Collaboration and Team Science[1] was launched in 2013 with the goal of advancing science by helping researchers put in place the infrastructure and processes needed to facilitate success in team-based science.

This chapter looks in turn at the funding and evaluation of team science. The final section presents conclusions and recommendations.

FUNDING FOR TEAM SCIENCE

A range of organizations fund team science. Examples of funders include (1) federal agencies, (2) private foundations and individual philanthropists, (3) corporations, (4) academic institutions that provide seed money or infrastructure, and (5) nonprofit organizations that obtain funding from private donors and/or the general public and use it to fund team science research (e.g., Stand Up to Cancer[2]). At a time of constrained public spending, alternative sources of funding become increasingly important to maintain the scientific enterprise. Additionally, a plurality of sources can potentially help to balance tensions between, for example, supporting an individual scientist to establish novel areas of research without "strings attached" versus more directed programmatic funding focusing on a specific research area (OECD, 2011). The wide range of funders and the evolving nature of their roles introduce many avenues through which funders can support and facilitate team science. In this section, we describe how funders can influence the conduct and support of team-based research, including a discussion of the broader context for the ways priorities are set.

Federal Funding for Team Science

Federal funding for team science has increased greatly over the past four decades. For example, agencies are increasingly providing funding to projects overseen by more than one principal investigator (PI). At NSF, the number of awards to projects with multiple PIs increased from fiscal year 2003 to fiscal year 2012, while the number of awards to individual PIs remained steady (National Science Foundation, 2013). At the National Institutes of Health (NIH), the number of multiple PI grants grew from 3 in 2006 (the first year such grants were awarded) to 1,098 in 2013, or 15–20 percent of all major grants funded (Stipelman et al., 2014). Agencies also

---

[1] The subcommittee is part of the Social, Economic, and Workforce Implications of IT and IT Workforce Development Coordinating Group within the National Information Technology Research and Development Program of the National Technology Council in the Executive Office of the President. See http://www.nitrd.gov/nitrdgroups/index.php?title=Social,_Economic,_and_Workforce_Implications_of_IT_and_IT_Workforce_Development_Coordinating_Group(SEW_CG)#title [May 2015].

[2] For more information, see http://www.standup2cancer.org/what_is_su2c [May 2015].

have increased their funding of research centers, which typically include multiple, related research projects that may be interdisciplinary and may involve industry or other stakeholders. For example, beginning in 1985 with a single center program, called the Engineering Research Centers, NSF created six more center programs over the following decade. By fiscal 2011, NSF invested nearly $298 million in these seven center programs, supporting 107 centers and engaging scientists at approximately 2,200 universities (National Science Foundation, 2012). At NIH, there were very few center grants until the mid-1980s, but the number of these grants to research centers and more loosely linked networks has increased steadily since then, as shown in Figure 9-1.

Prioritizing Research Topics and Approaches

Public and private funders work closely with both the scientific community and policy makers to establish research priorities and approaches. Federal agencies that fund research are led and staffed by scientists, convene scientific advisory bodies (e.g., the Department of Energy's High Energy Physics Advisory Panel), and allocate funding through peer review by panels of scientists. Major new federal research programs often involve

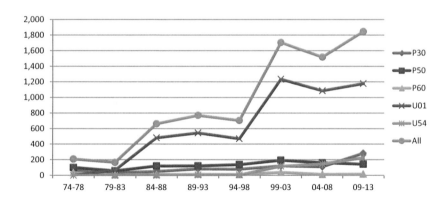

FIGURE 9-1 Number of commonly awarded National Institutes of Health Center/ Network Grants by mechanism over time.
NOTES:
P30 = Center Core Grants
P50 = Specialized Center
P60 = Comprehensive Center
U01 = Research Project Cooperative Agreement
U54 = Specialized Center Cooperative Agreement
SOURCE: Unpublished data provided by the National Institutes of Health.

years of engagement and discussion among funding agencies, the scientific community, policy makers, and other stakeholders. For example, the 1990 congressional mandate for the U.S. Global Climate Change Research Program emerged from an array of "bottom-up" research projects initiated by scientists (Shaman et al., 2013). According to Braun (1998, p. 808), if they are strategic, then "funding agencies are in a good position to balance demands from both the political and scientific sides."

Through this collaborative process of setting research priorities, federal agencies have increasingly supported team science approaches in recent years (see further discussion below). Nevertheless, some scientists fear that increased public funding of large groups of scientists focusing on particular topics transfers too much control of research topics, approaches, and goals away from the scientific community and to bureaucrats (e.g., Petsko, 2009). Such views reflect the traditional role of individual investigators and professional societies (most of which are discipline based) in setting research agendas through publications, meetings, annual conferences, and peer review panels. However, as the number of scientific specializations increases and the public and policy makers seek solutions to scientific and societal problems, the scientific enterprise can benefit when funders look across disciplines or individual studies within a discipline to see the "big picture" of research needs and opportunities. Critiques of the peer review processes used in awarding research grants as too conservative (National Institutes of Health, 2007; *Nature*, 2007; Alberts et al., 2014) reinforce the potential benefit if funders consider whether new research areas need to be stimulated. In some instances, if scientists continue to focus on already well-explored problems or approaches that hold limited potential to add to existing scientific knowledge, then funders may need to set new directions and priorities (Braun, 1998). Furthermore, given the historically individual- and discipline-based incentive structure of academia and scientific journals, funders are positioned to provide incentives for alternatives to these approaches.

The Growing Role of Private Funders

Individual philanthropists and private foundations are beginning to play a larger role in establishing research priorities, and the continued debate regarding how much funders should influence the directions of science extends to these private entities. A policy analyst at the American Association for the Advancement of Science recently commented: "For better or for worse, the practice of science in the 21st century is becoming shaped less by national priorities or by peer review groups and more by the particular preferences of individuals with huge amounts of money" (Broad, 2014).

Philanthropic giving influences scientific research through investments such as establishing new institutes or providing funds through universities. Private foundations and wealthy individuals contribute an estimated $7 billion per year to research conducted at U.S. universities, with a strong emphasis on translational medical research (Murray, 2012). They typically target particular scientific areas or address specific societal problems, prioritizing research topics narrowly, rather than in broad strategic ways.

The growth of private funding raises questions for federal policy makers and research funding agencies. One concern is that scientists funded by philanthropists with particular research agendas, who also sit on funding agency advisory panels and peer review panels, may have a significant influence over the priorities set by federal funding agencies. Another is that wealthy individuals may ignore important fields of science that lie outside their direct interests.

Funding Models, Funding Mechanisms, and Organizational Structures

Whether funding individual or team science, once funders establish research needs and priorities, they develop funding models and mechanisms to address the identified needs. Funding organizations differ widely in their models and mechanisms of funding (e.g., Stokols et al., 2010). *Funding models* are generic mechanisms for funding science (e.g., grants, prizes, donations), while *funding mechanisms* are specifically targeted incarnations of funding models. For example, NIH P50 is a specific type of funding mechanism to support research centers, and the Google Lunar XPrize is a specific competition inviting private teams to land a robot safely on the surface of the moon. Some funders are experimenting with "open" funding mechanisms. For example, the Open Source Science Project (2008–2014) used a web-based micro financing approach to support individual or team research, and the Harvard Medical School used an open funding mechanism to generate research topics on Type I diabetes (Guinan, Boudreau, and Lakhani, 2013).

As noted in previous chapters, these various funding mechanisms may support various *organizational structures* for team science (Hall et al., 2012c), ranging from small science teams to global networks The amount of funding often dictates the magnitude and, thereby, complexity of the organizational structure; a worldwide research network requires more resources than a university research center, which requires more resources than a single research project. In addition, science teams or larger groups may be funded by multiple public and private sources. For example, the NSF investment in Science and Technology Centers discussed in the previous chapter is multiplied by funding from industry and universities, while the Koch Center for Integrative Cancer Research at Massachusetts Institute of Technology combines private donations with university funding and federal support as a National Cancer Institute-designated Cancer Research Center. In addition to important differences in how team science is funded, there is wide variation in how the research funding can be used. Common expenses include academic salaries, student tuition, equipment, materials, and space. Some budgets permit funds for training (e.g., cross-training for interdisciplinary teams), core units (e.g., administrative or statistical support), discretionary developmental projects (e.g., small mid-course pilot projects), travel for collaborators, and conference attendance. These variations in how funding can be used have important implications for team science funding, raising questions about how and when funders might provide support for:

- Planning or meeting grants to support the developmental phases of team science, which provides an incubator space to generate or advance new cross-disciplinary ideas (National Research Council, 2008; Hall et al., 2012a, 2012c).
- Travel funds to enable geographically dispersed teams to meet face-to-face, which can enhance communication and trust (National Research Council, 2008; Gehlert et al., 2014), as discussed in Chapter 7.
- Developmental or pilot project funds to enable flexible funds for just-in-time innovations or new integrative ideas that emerge during larger collaborative projects (Hall et al., 2012a; Vogel et al., 2014).
- Professional development funds, which can be used to promote the early development of collaborations and facilitate team processes that enhance effectiveness (see Chapter 5).
- Flexible funds for leaders of team science projects to allow them to make "real-time" adjustments for projects as project needs unfold. For example, leaders might be allowed to move funds between subprojects, adjust the timing of funding plans, and/or provide incentives and rewards for successful team research (National Cancer Institute, 2012).

Agencies often use public announcements, referred to as Funding Opportunity Announcements (FOAs), Program Solicitations, or Program Announcements to emphasize scientific priorities and influence the particular approaches used to implement those priorities. Additionally, these announcements delineate the type of mechanism and describe the intended organizational structure for supporting that approach. Language in the FOAs can encourage or stipulate particular approaches for conducting science (e.g., interdisciplinary, transdisciplinary, translational) and organizational configurations (e.g., centers or teams; see Table 9-1). For example, the program solicitation for NSF's CyberSEES Program states, "Due to this program's focus on interdisciplinary, collaborative research, a minimum of two collaborating investigators (PIs/Co-PIs) is required" (National Science Foundation, 2014a).

However, agency leaders and staff experience a tension between providing clear guidance (which may become too prescriptive) and encouraging flexible responses from scientists, based on their particular research contexts and capabilities. In addition, agency employees sometimes lack understanding of team science processes and outcomes. As a result, they sometimes develop public announcements that include vague language about the type of collaboration and the level of knowledge integration they seek in the desired research[3] (see Table 9-1). Announcements may lack sufficient guidance to facilitate interaction (e.g., by specifying the timing and frequency of in-person or virtual meetings or the inclusion of professional development plans). If the funder is soliciting interdisciplinary or transdisciplinary proposals, then these announcements may lack sufficient guidance to facilitate the deep knowledge integration that is required to carry out such research.

When funders do clearly articulate their goals for team science, they provide signals to scientists and institutions, which can in turn help facilitate culture change in the broader scientific community. For example, an earlier National Academies study reported that many scientists would like research universities to recognize and reward interdisciplinary research (National Academy of Sciences, National Academy of Engineering, and Institute of Medicine, 2005). In response to signals from NIH, the promotion and tenure guidelines for the University of Virginia School of Medicine support such recognition and reward. The guidelines include the statement, "The NIH roadmap for patient-oriented research endorsed team science and established the expectation of expertise for interdisciplinary investigation and collaboration" (Hall et al., 2013). This language reflects the medical school's effort to align its institutional rewards and incentives with team-based approaches to conducting science and highlights the important role that funding agencies can play in influencing the scientific community.

EVALUATION OF TEAM SCIENCE

Funders evaluate science teams and larger groups throughout the evolution of a research endeavor, beginning with the proposal review, then dur-

---

[3] Professional leadership development to increase agency employees' understanding of team science, as recommended in Chapter 6, could help improve the clarity of communication in research solicitations involving team science.

TABLE 9-1 Examples of Federal Research Funding Announcements That Support Team Science

Agency	FOA/Program Soliciation Number	Program Name	Type of Mechanism	Organizational Structure	Funds	Examples of Language Related to Team Science
NSF	NSF 07-558	Engineering Virtual Organization Grants	Standard grant	Seed money to create engineering virtual organization.	**Small** $2M total for ~10–15 awards	"EVOs extend beyond small collaborations and individual departments or institutions to encompass wide-ranging, geographically dispersed activities and groups."
DoE	DE-FOA-0000919	Collaborative Research in Support of GOAmazon Campaign Science	Research grant award	Projects will be affiliated with the multilateral campaign. Investigators are expected to coordinate their research with other investigators and with reps of the ARM Climate Research Facility.	**Small** $2.3M total for ~6-8 awards ($50k–350k per award)	"Emphasis on collaboration; those involved must be truly collaborative in the conduct of research, including definition of goals, approach, and work plan."
NSF	NSF 13-500	Cyber-Enabled Sustainability Science and Engineering (CyberSEES)	Standard grant	Team must include at least two investigators from distinct disciplines.	**Small to Medium** $12M total for ~12–20 awards	"Team composition must be synergistic and interdisciplinary"; "focus on interdisciplinary, collaborative research; a minimum of two

(continued) collaborating investigators (PIs/Co-PIs) working in different disciplines is required."

NSF	NSF 12-011	CREATIV: Creative Research Award for Transformative Interdisciplinary Ventures	New grant mechanism for special projects	Any NSF support topic area: interdisciplinary, high-risk, novel, potentially transformative.	**Medium** Up to $1M total up to 5 years	"Must integrate across multiple disciplines;" "the proposal must identify and justify how the project is interdisciplinary;" "encourage cross-disciplinary science;" "break down any disciplinary barriers;" "proposals must be interdisciplinary."
NIH	RFA-AG-14-004	Roybal Centers for Translational Research on Aging	P30	Center organized around thematic area—includes translational research activities, pilot projects, cores and coordination center (optional).	**Medium** $3.9M/year ~8–12 awards	"To galvanize scientists at several academic institutions;" "of particular interest are projects that incorporate approaches from emerging interdisciplinary areas of behavioral and social science, including behavioral economics; the social, behavioral, cognitive and affective neurosciences; neuroeconomics; behavior genetics and genomics; and social network analysis."

Continued

206

TABLE 9-1 Continued

Agency	FOA/ Program Solicitation Number	Program Name	Type of Mechanism	Organizational Structure	Funds	Examples of Language Related to Team Science
NASA	NASA ROSES A.11, NNH13Z-DA001N-OVWST	Ocean Vector Winds Science Team	Standard grant	Research that requires vector wind and back-scatter measure-ments provided by QuikSCAT and the combined QuikSCAT/Mi-dori-2 SeaWinds scatterometers.	$4.5M/year, expected ~25 awards	"Oceanographic, meteo-rological, climate, and/or interdisciplinary research."

NOTES:
DoE = U.S. Department of Energy
FOA = Funding Opportunity Announcements
NASA = National Aeronautics and Space Administration
NSF = National Science Foundation
RFA = Request for Applications

ing the active research project, and finally following the end of the formal grant period.

Proposal Review

Once funders have mechanisms in place to support team science, they must solicit and facilitate the review of proposals submitted for funding. Sometimes this review process involves internal review by program officers, but more often it involves peer review by experts in the field of study (Holbrook, 2010). There are a number of challenges that arise when reviewing team science proposals, especially when the research is interdisciplinary in nature. Challenges include issues such as composition of review panels and needed scientific expertise (Holbrook, 2013). To adequately review an interdisciplinary or transdisciplinary proposal, funders need to identify and recruit reviewers with expertise in the range of disciplines and methods included (Perper, 1989). It is often not sufficient, however, to have the specific expertise related to the elements of a proposal, as individuals with specialized expertise may not have sufficient breadth of knowledge or perspective to evaluate the integration and interaction of disciplinary or methodological contributions of an interdisciplinary proposal.

This may be particularly relevant in the case of agencies such as NIH where reviewers have been increasingly more junior (Alberts et al., 2014; *Nature*, 2014). Less experienced reviewers especially need review criteria to be clear, including what is being judged and how quality is defined (Holbrook and Frodeman, 2011; also see National Science Foundation, 2011, for a description of the agency's merit review criteria).

In a recent empirical study of the grant proposal process at a leading research university, Boudreau et al. (2014) lent support to the view that peer reviewers may be too conservative. The authors found that members of peer review panels systematically gave lower scores to research proposals closer to their own areas of expertise and to highly novel research proposals. They suggested that, if funders wish to support novel research, then they prime reviewers with information about the need for and value of novel research approaches in advance of the review meeting. A related concern is that some reviewers from individual disciplines may be biased against interdisciplinary research, potentially complicating the evaluation of the science itself (Holbrook, 2013). NSF's Workshop on Interdisciplinary Standards for Systematic Qualitative Research (Lamont and White, 2005) produced an approach for establishing review criteria that could be applied to interdisciplinary research more broadly. Furthermore, review panels for some cross-disciplinary translational research, such as the patient-centered outcomes research funded by the congressionally mandated Patient-Centered Outcomes Research Institute, includes non-scientist reviewers (at least two on each review panel). Such stakeholders are included to "help ensure the research . . . reflects the interests and views of patients" (Patient-Centered Outcomes Research Institute, 2014). The creation of the institute and inclusion of these stakeholders is an indication that patient advocacy groups are influencing biomedical research and health care practice (Epstein, 2011).

A number of additional issues can arise in the process of reviewing proposals. For example, involving many institutions may strengthen a given team science project (e.g., by bringing more resources or perspectives to the project), but this can potentially create a bias in favor of having more institutions. Reviewers may rate proposals including multiple institutions more favorably than those including fewer institutions, rather than basing their ratings entirely on scientific merit (Cummings and Kiesler, 2007). In some cases, moreover, reviewers from an institution included in a proposal must excuse themselves from review in order to avoid conflict of interest (e.g., in NIH and NSF panel reviews); the larger the proposed science group, the higher the likelihood that review members will need to leave the room. As a result, with larger and more complex projects, relatively fewer panel reviewers will remain in the room to judge the proposals. Such complications have prompted changes in agency policies for managing conflict-of-interest issues in the peer review process. For example, NIH (2011) issued a revised review policy based on "the increasingly multi-disciplinary and collaborative nature of biomedical and behavioral research."

As discussed in the previous chapter, larger and more complex projects are also at greater risk for collaborative challenges after funding, yet there are typically no sections of the grant application devoted to describing management or collaboration plans. Review criteria are typically focused on the technical and scientific merit of the application, and not the potential of the team to collaborate effectively. The Trans-agency Subcommittee on Collaboration and Team Science mentioned above believes that including collaboration plans in proposals will help ensure that the needed infrastructure and processes are in place. The subcommittee has engaged in a series of workshops and projects specifically to develop guidance for (a) researchers, including key components to consider when developing collaboration/ management plans; (b) agencies, including potential language for program officers to use when soliciting collaboration plans from investigators or guidance to researchers; and (c) reviewers, including evaluation criteria for reviewers of collaboration plans of submitted by investigators as part of a funding proposal.

Team charters typically outline a team's direction, role, and operational processes, whereas agreements or contracts outline specific terms that multiple parties formally or informally establish verbally or in writing. The use of charters and agreements for addressing specific collaborative factors such as conflict, communication, and leadership has been discussed in the literature (e.g., Shrum, Gernuth, and Chompalov, 2007; Bennett, Gadlin, and Levine-Finley, 2010; Asencio et al., 2012; Bennett and Gadlin, 2012; Kozlowski and Bell, 2012). Importantly, as noted in Chapter 7, Mathieu and Rapp (2009) showed that the use of charters increased team performance, and that the quality of the charter mattered. In a study by Shrum, Genuth, and Chompalov (2007), the greater the number of participants, teams, and organizations included in a large, multi-institutional research project, the more frequently formal contracts were used. Although two-thirds of the collaborations studied by Shrum, Genuth, and Chompalov (2007) used some form of formal contract, the contracts were often very specific (e.g., to specify roles and assignments or rules for reporting developments within/outside of the collaboration) or were not drawn up until the end of the project.

Collaboration plans, as described here, build upon the goals of charters and agreements/contracts, but provide a broader framework to help address the breadth of issues outlined in this report. The plans include the use of charters, agreements, and contracts to achieve specific objectives. A study (Woolley et al., 2008) examining the influence of collaboration planning demonstrated that (p. 367) "team analytic work is accomplished most effectively when teams include task-relevant experts and the team explicitly explores strategies for coordinating and integrating members' work." The authors found that high task expertise in the absence of explicit collaborative planning actually decreased team performance.

This report has highlighted evidence related to factors at many levels that influence the effectiveness of team science. The primary goal of collaboration plans is to engage teams and groups in formally considering the various relevant factors that may influence their effectiveness and deliberately and explicitly planning actions that can help maximize their effectiveness and research productivity. Collaboration plans can serve to provide a framework for systematically considering the primary domains covered in this report. Federal agencies have begun requiring plans such as data management plans (e.g., NSF[4]) or leadership plans (e.g., NIH[5]), which contain elements of collaboration plans. However, these required plans are designed for more specific purposes or for specific mechanisms.

Emerging guidelines for broader collaboration plans, developed by the trans-agency subcommittee, would require proposals to address 10 key aspects of the proposed project: (1) Rationale for Team Approach and Team Configuration; (2) Collaboration Readiness (at the individual, team, and institutional levels); (3) Technological Readiness; (4) Team Functioning; (5) Communication and Coordination; (6) Leadership, Management, and

---

[4] See http://www.nsf.gov/bio/pubs/BIODMP061511.pdf [May 2015].
[5] See http://grants.nih.gov/grants/guide/notice-files/NOT-OD-07-017.html [May 2015].

Administration; (7) Conflict Prevention and Management; (8) Training; (9) Quality Improvement Activities; and (10) Budget/Resource Allocation (Hall, Crowston, and Vogel, 2014). Collaboration plans should vary in relation to the size and complexity of the scientific endeavor and take into account unique circumstances of the proposed team or group. The goal is to effectively collaborate to more rapidly advance science.

Program Evaluation

Evaluation approaches include *formative evaluation,* which provides ongoing feedback for project improvement (D. Gray, 2008; Vogel et al., 2014) and retrospective *summative evaluation,* which provides lessons for enhancing future programs (e.g., Institute of Medicine, 2013; Vogel et al., 2014). Public or private funders may require one or both types of evaluation as a condition of funding (Vogel et al., 2014) or conduct or commission evaluations on an ad hoc basis (e.g., Chubin et al., 2009). The complexities introduced by team-based research need to be considered when developing a comprehensive evaluation plan. However, a recent review of more than 60 evaluations of NIH center and network projects from the past three decades found that while a majority of evaluation studies included some type of evaluation of the research process, this important dimension often was represented with either a single variable or a limited set of variables that were not linked to one another or to program outcomes in any conceptually meaningful way (The Madrillon Group, 2010).

Improvement-Oriented Approaches

Improvement-oriented or formative evaluation aims to enhance the ongoing management and conduct of a project by providing feedback to support learning and improvement (e.g., D. Gray, 2008; The Madrillon Group, 2010). This can be done in a number of different ways, including embedding evaluators within the team or group (e.g., D. Gray, 2008), engaging team science researchers to study the projects (e.g., Cummings and Kiesler, 2007), and collaborating with science of team science scholars or evaluators at a federal agency (e.g., Porter et al., 2007; The Madrillon Group, 2010; Hall et al., 2012b). For larger and longer-duration projects, especially university-based research centers, it is not unusual for a funding agency to conduct a site visit in which program officers visit the principal investigators (PIs) and have in-person discussions with project participants. Site visits allow funders to learn about the people involved in the projects, the research being conducted, and any barriers or hurdles being encountered.

Outcome-Oriented Approaches

Increased funding of team science has raised questions within the scientific community about the effectiveness of team approaches relative to more traditional, solo science, which has put pressure on funders to demonstrate the value of their investments through *summative evaluation* of outcomes (Croyle, 2008, 2012). Whether conducted as a case study or to compare what the project has achieved with a known benchmark or standard, summative evaluation can provide valuable information to funders and other stakeholders in the scientific community (Scriven, 1967). However, evaluating the outcomes of team science projects can be difficult, as discussed in Chapter 2. For example, the goals of small teams may entail the creation and dissemination of new scientific knowledge, but the goals of larger groups may include translation of scientific knowledge into new technologies, policies, and/or community interventions. Thus, the first step toward evaluating outcomes is to clearly specify all desired outcomes from the beginning. For example, if translation is a desired outcome, then Funding Opportunity Announcements (FOAs) could provide examples of outputs from research projects that synthesize and translate research findings into formats useful for a variety of stakeholder groups; such outputs might include written briefs or informational videos for use in clinical practice or new product development.

A summative evaluation can be completed by researchers themselves (e.g., through a final report or published journal article), by program evaluators contracted by funding agencies, by internal agency staff in collaboration with grantees, or by team science researchers. In all cases, the purpose is to establish lessons learned for the development and implementation of subsequent science teams, larger groups, or research programs (Hall et al., 2012b; Vogel et al., 2014). There are many dimensions to choose from when conducting an evaluation of team science outcomes, including identifying or developing metrics of outputs (e.g., publications, citations, training; see Wagner et al., 2011 for a discussion of interdisciplinary metrics), and identifying the intended targets of these outputs (research findings may be targeted to academics, business, or the general public; see Jordan, 2010, 2013). In addition, the evaluator must consider the type of innovation sought by the project (e.g., incremental or small improvements vs. radical or discontinuous leaps; see Mote, Jordan, and Hage, 2007), the time frame (e.g., short-term vs. long-term outcomes), and the type of intended long-term impact (e.g., science indicators; see Feller, Gamota, and Valdez, 2003). Evaluators can also use a range of methods to judge how successful particular team science projects have been, such as citation analysis and the use of quasi-experimental comparison samples and research designs (Hall et al., 2012b).

As discussed in Chapter 2, evaluators have tended to rely on publication data (bibliometrics) as metrics of the outputs and outcomes of team science. While funders and evaluators recognize the need for new metrics to capture broader impacts, such as improvements in public health (Trochim et al., 2008), developing methodologically and fiscally feasible metrics has proven difficult (see Chapters 2 and 3). Other challenges to conducting a thorough evaluation arise due to unavailability of data from a range of programs and projects. In addition, little research to date has used experimental designs, comparing team science approaches or interventions[6] with control groups to identify impacts.

The recent development of "altmetrics" provides helpful data that may be used to improve evaluation of team science projects (Priem, 2013; Sample, 2013). In 2010, a group of scientists called for consideration of all products of research grants rather than just peer-reviewed publications, including sharing of raw data and self-published results on the web and through social media; they also called for development of "crowdsourced" automated metrics tied to the products, such as reach of Twitter posts or blog views (Priem et al., 2010). The new movement already has had some effects, as NSF has changed the language of required biosketches to include products such as datasets, software, patents, and copyrights. Piwowar (2013) contended that altmetrics give a fuller picture of how the products of scientific research have influenced conversation, thought, and behavior.

As emphasized in this report, it is important to evaluate the team science processes and to study the relationships of these processes to research outcomes and impacts in order to understand potential mediators and moderators of successful team science outcomes. By doing so, funders can contribute to the knowledge needed to develop evidence-based support for team science. In addition, studies can examine not only the relationships between outcomes and particular funding mechanisms (Druss and Marcus, 2005; Hall et al., 2012b) but also outcomes and measures of team processes (e.g., The Madrillon Group, 2010, Stipelman et al., 2010) to increase the knowledge base and enhance funders' ability to better support team science.

In a time of federal budget constraints, funding agencies are becoming increasingly aware of the potential advantages of using systematic and scientific approaches to managing, administrating, and setting priorities and allocating funds. The Office of Management and Budget in the Executive Office of the President (2013) released a government-wide memo that calls for using evidence and innovation to improve government performance.

---

[6] As noted in Chapter 6, an ongoing study by Salazar and colleagues uses an experimental design to test interventions designed to facilitate knowledge integration in interdisciplinary and transdisciplinary projects (see http://www.nsf.gov/awardsearch/showAward?AWD_ID=1 262745&HistoricalAwards=false [May 2015]).

The memo emphasizes (p. 3) "high-quality, low-cost evaluations and rapid, iterative experiments" and the use of "innovative outcome-focused grant designs." Agencies have begun responding to this message. For instance, a recent report by NIH (2013) summarized and recommended:

> . . . ways to strengthen NIH's ability to identify and assess the outcomes of its work so that NIH can more effectively determine the value of its activities, communicate the results of studies assessing value, ensure continued accountability, and further strengthen processes for setting priorities and allocating funds.

The Office of Management and Budget memo and the NIH report highlight the need for the development of more evidence-based strategies to facilitate and support team science. The science of team science community is well poised to help address these issues.

SUMMARY, CONCLUSIONS, AND RECOMMENDATIONS

Many public and private organizations fund and evaluate team science. Public and private funders typically use a collaborative process to set research priorities, engaging with the scientific community, policy makers, and other stakeholders. Hence, they are well-positioned to work with the scientific culture to support those who want to undertake team science. When soliciting proposals for team science, federal agency staff members sometimes write funding announcements that are vague about the type and level of collaboration being sought. At the same time, the peer review process used to evaluate proposals typically focuses on technical and scientific merit, and not the potential of the team to collaborate effectively. Including collaboration plans in proposals, along with guidance to reviewers about how to evaluate such plans, would help ensure that projects include infrastructure and processes that enhance team science effectiveness. The committee's review of research and practice on funding and evaluation of team science in this Chapter raises several important unanswered questions, which are discussed in Chapter 10.

CONCLUSION. *Public and private funders are in the position to foster a culture within the scientific community that supports those who want to undertake team science, not only through funding, but also through white papers, training workshops, and other approaches.*

RECOMMENDATION 7: **Funders should work with the scientific community to encourage the development and implementation of new collaborative models, such as research networks and consortia; new team science incentives, such as academic rewards for team-based research** (see Recommendation #6); and resources (e.g., online repositories of information on improving the effectiveness of team science and training modules).

CONCLUSION. *Funding agencies are inconsistent in balancing their focus on scientific merit with their consideration of how teams and larger groups are going to execute the work (collaborative merit). The Funding Opportunity Announcements they use to solicit team science proposals often include vague language about the type of collaboration and the level of knowledge integration they seek in proposed research.*

RECOMMENDATION 8: Funders should require proposals for team-based research to present collaboration plans and provide guidance to scientists for the inclusion of these plans in their proposals, as well as guidance and criteria for reviewers' evaluation of these plans. Funders should also require authors of proposals for interdisciplinary or transdisciplinary research projects to specify how they will integrate disciplinary perspectives and methods throughout the life of the research project.

Part IV

A Path Forward

10

Advancing Research on the Effectiveness of Team Science

The committee's review of the research related to the study charge yielded many new insights into approaches to enhance the effectiveness of team science. However, it also identified gaps in the evidence base where further research is needed. Here, we discuss some of the research needs in greater detail and the promise of new methods for use in addressing them.

TEAM PROCESSES AND EFFECTIVENESS

As discussed in Chapters 1 and 2, science teams and larger groups share many challenges with teams in other contexts, and, thus, the research on teams in other contexts is relevant for team science. In particular, team process factors, such as the development of shared understanding of team goals and roles, have been shown to influence the ability of teams to achieve their goals, both in science and in other contexts. Drawing on this research, previous chapters recommended actions and interventions in three aspects of team science—composition, professional development, and leadership. At the same time, however, we have noted the need for further "basic" research on team processes within science teams and larger groups and how these processes are related to scientific discovery and translation.

Improving an understanding of the processes of team science will require interdisciplinary collaboration involving experts in the various disciplines that study teams and organizations (i.e., psychology, organizational behavior, communications) and in the science of team science and related fields (such as economics, science policy, philosophy of science and systems science), along with team science practitioners. Investigators working together could develop a comprehensive, multi-method measurement approach to investigating the dynamics and outcomes of science teams and larger groups. Such an approach includes, but is not limited to, bibliometric indices, co-authorship network analyses, experts' subjective appraisals of team science processes and products, and surveys and interviews of team science participants. In particular, valid and reliable metrics are needed to more clearly understand the process of deep interdisciplinary knowledge integration and how it varies in unidisciplinary, multidisciplinary, interdisciplinary, and transdisciplinary science teams and groups (Wagner et al., 2011). Along with advances to metrics, investigators working together can apply rigorous experimental methodology (e.g., manipulations, control conditions, before–after data) to science teams and groups to develop a deeper understanding of causal mechanisms underlying effective team science.

Future efforts to understand team science processes can be aided by new approaches, such as the complex adaptive system approach discussed in Chapter 2. In addition, new data collection methods are becoming available, such as the use of wearable electronic badges that unobtrusively trace scientists' interactions as they work (see further discussion below). This research should use methods sophisticated enough to address longitudinal changes across levels of analysis (e.g., individual, team, organizational) and the resulting mediators and moderators of the hypothesized effects; such methods are described in the final section of this chapter.

Specific research gaps associated with science team composition, professional development, and leadership are highlighted in the following three subsections.

Team Composition and Assembly

In Chapter 4, we concluded that methods and tools that allow practitioners to consider team composition systematically appear promising and recommended that those involved in assembling science teams and larger groups apply these methods and tools. As team science leaders begin to apply task analytic methods to compose science teams and larger groups (implementing Recommendation #1 in the Summary), evaluation studies are needed to guide refinements and improvements to these applications. An ongoing cycle of implementation, evaluation, and revision would further strengthen the ability of team and group leaders to identify the task-relevant diversity needed to achieve the scientific or translational goals of the project. Chapter 4 also discussed recent research on the team assembly process. Further research on the assembly process in science teams, including comparative studies of the processes and outcomes of self-assembled versus assigned teams, would provide valuable information to the scientific community, funding agencies, and university administrators. Studies on the implementation and impacts of the new research networking tools that are being adopted by many research institutions would also be valuable.

At the same time, Chapter 4 highlighted the disagreements and uncertainties in the research to date about how various individual characteristics may affect team outcomes. In light of these uncertainties, there is a clear need for further and more sophisticated research on how the multiple individual characteristics of the team or group members combine within science teams and groups, and how these interactions and processes are related to effectiveness. This research would address such questions as:

- What is the role of individual characteristics (including dispositional qualities such as social intelligence) in team processeses and effectiveness?
- How do the interactions among subgroups (whose members may share multiple similar characteristics) affect team processes and effectiveness?
- How does team composition interact with team processes to influence team effectiveness?
- How do changes in science team or group membership impact team processes and outcomes?
- How may the various roles team or group members play (e.g., connectors/brokers, leaders, scientists with particular expertise, community stakeholders) be characterized? What are the interrelationships between these roles, and how do they affect team processes and effectiveness?

Professional Development and Education for Team Science

In Chapter 5, we concluded that several types of professional development show promise to improve the processes and outcomes of science teams. As universities, researchers, and practitioners begin to create professional development opportunities for science teams (implementing Recommendation #2 in the Summary), ongoing evaluation of these opportunities would provide valuable information for continuous improvement of them. In addition, more basic research on how science teams and groups learn and develop would enhance future professional development.

We also concluded that colleges and universities are increasingly developing cross-disciplinary programs designed to prepare students for team science, but that little is known about the effectiveness of these programs. In particular, we noted that some of these programs do not clearly articulate the competencies they are intended to develop and they target a variety of competencies. The literature has produced a plethora of competencies that overlap to some degree and also have differences. And, little empirical research is available on the effectiveness of such programs in developing the various competencies that they target. Methods used to date to evaluate these programs rely heavily on case studies and expert reviews.

Addressing these gaps in the research evidence will require collaboration between the multiple communities engaged in interdisciplinary education and the team-training research community. Through such collaboration, researchers could create methods for assessing both collaborative and intellectual outcomes to identify core competencies that could then be systematically integrated into graduate and undergraduate programs to prepare students and team members for team science. More generally, collaboration among these communities would make it possible to conduct more robust prospective studies that compare and explicitly evaluate the relative effectiveness of various educational programs designed to prepare students for team science. In particular, it will be important to address the following unanswered questions:

- How is variation in the competencies developed through education and/or professional development related to team science processes and outcomes? For example, under what conditions do teamwork training (focused on team-related knowledge and skills) and task work training (focused on the scientific knowledge and skills) enhance scientific productivity?
- What educational or professional development approaches are most effective in developing the targeted competencies at different educational and career levels (e.g., doctoral education vs. senior investigator)?

Team Science Leadership

In Chapter 6, we concluded that 50 years of research on team and organizational leadership in contexts other than science provides a robust foundation of evidence to guide creation of leadership development programs for team science leaders. As researchers and team science practitioners begin to develop such programs (implementing Recommendation #3 in the Summary), ongoing evaluation is needed to inform continued revisions and improvements. An ongoing cycle of continuous improvement, based on testing and evaluating the new courses, would enhance the quality of future leadership development programs for team science. Such efforts would enhance participants' capacity to lead in ways that facilitate positive team processes and enhance scientific and translational effectiveness. At the same time, more basic research could guide these efforts by, for example, investigating the applicability of promising recent leadership approaches to science teams and larger groups, including contextual leadership, emergent leadership, team leadership, and shared leadership.

SUPPORT FOR VIRTUAL COLLABORATION

In Chapter 7, we concluded that when scientific colleagues are geographically remote from one another, issues such as lack of shared vocabularies and experiences and role confusion may be exacerbated relative to face-to-face teams or groups. Although the research supports our recommendation that team leaders take several steps to address these issues, it would be valuable to conduct further research on the extent to which the research on teams and groups and principles for effectiveness identified in Chapters 3 through 6 are applicable to virtual science teams and larger groups.

We also concluded that technology for virtual collaboration often is designed without a true understanding of users' needs and limitations and may thus impede such collaboration. Hence, further research is needed to evaluate how tools and practices for virtual collaboration affect team processes and outcomes. This requires that researchers, technology developers, and technology users work together to conduct research on user-centered design and human-systems integration so that the various tools for collaboration are interoperable and are aligned with users' activities and capabilities.

INSTITUTIONAL AND ORGANIZATIONAL SUPPORT FOR TEAM SCIENCE

In Chapter 8, we observed that many universities are launching efforts to promote and support interdisciplinary team science, but research is sorely needed to guide these efforts, so that they succeed in fostering team science and advancing scientific discovery and translation. To date, the impact of these efforts on the amount and quality of team science research remains to be systematically evaluated. In particular, we noted that university-industry research collaborations have grown faster than the knowledge of how to manage them effectively. Limited systematic, rigorous research is available on such partnerships, and there is a dearth of research on failed collaborations. In addition, we noted that research on the relationship between design of the built environment and scientific collaboration remains theoretically debated and empirically mixed. Some studies have found a positive relationship between spatial proximity and scientific collaboration, but additional research is needed to improve understanding of the relationship between the design of the built environment and team science effectiveness. A broader focus for this research would examine cultural and social factors intertwined with the spatial environment that may jointly affect collaborative processes and outcomes.

A few studies are beginning to examine some specific university strategies to promote interdisciplinary team science. For example, one recent study examined how Harvard Medical School's "open" call for research ideas aided development of research topics on Type I diabetes (Guinan, Boudreau, and Lakhani, 2013). The committee encourages more agencies and universities to study and learn from existing and emerging strategies to enhance the way science is supported and conducted.

A follow-on study to the 2005 National Academies survey of institutions and individuals conducting interdisciplinary research (National Academy of Sciences, National Academy of Engineering, and Institute of Medicine, 2005) might be a helpful step in guiding university efforts.[1] The findings would illuminate what progress has been made in the past decade, what obstacles still remain, and what research-based promising practices can be identified. In turn, the results of this new, follow-up study could be used in formulating more specific research studies to increase understanding of how alternative types of organizational and institutional policies and practices affect team science.

More generally, research on university efforts would provide greater clarity if it included more field and quasi-experimental study designs with longitudinal and panel components to examine the outcomes of university efforts over time. Studies of university-industry partnerships and other multi-stakeholder team science projects are needed to examine choices of institutional partners, factors related to both success and failure of these projects, formal and informal management practices, and the nature of their institutional impacts. Such studies would benefit from the development of data collection strategies and a performance data system that is transparent, meaningful, and accessible to researchers.

In Chapter 8, we also noted a few, isolated examples of university efforts to change policies and practices related to awarding credit for team science in the promotion and tenure process. Despite such exceptions, university policies for promotion and tenure review typically do not provide comprehensive, clearly articulated criteria for evaluating individual contributions to team-based research. Recognizing that disciplines, departments, and universities will continue to establish and apply their own criteria for evaluating research contributions, we recommended that universities and disciplinary societies proactively develop broad principles for assigning individual credit for team-based work. Targeted research is needed to inform these efforts, along with research on the feasibility and effectiveness of providing team rewards (e.g., bonuses, public recognition) for team-based work.

---

[1] Such a study need not be conducted by the National Academies.

More generally, research is needed to increase understanding of the promotion and tenure process as it relates to team science. A valuable first step would be a systematic survey of U.S. universities' promotion and tenure policies related to evaluating individual contributions to team-based research. The limited information currently available suggests that such policies include a relatively narrow range of criteria relative to the broad range of potential meaningful contributions an individual can make to a science team. Further research is needed to develop evidence-based principles for evaluating contributions such as being a "broker" who brings individuals and/or organizations together (a role that has been shown to facilitate innovation as discussed in Chapter 4).

In addition, research is needed to understand how such new principles and criteria could best be implemented, addressing such questions as:

- To what extent are university-wide policies implemented and adhered to?
- What factors, such as university, school, or departmental leadership and culture, influence the uptake of new policies?
- How long does it typically require before policy changes affect practice within promotion and tenure committees?

Research is also needed to explore team rewards for team science. Although many members of science teams and groups work at universities, others are found in industrial research and development laboratories, freestanding science facilities (e.g., particle accelerators or large observatories), federal laboratories, and public and private research centers and institutes. Regardless of where they are employed, scientists and other stakeholders engaged in collaborative research may respond to incentives and rewards provided by their employers. To date, despite the rapid growth of teams in science and other sectors of the economy, organizational incentive systems have focused primarily on rewarding individual achievements. Further research is needed to develop and test team-based rewards for team-based accomplishments. Such research would benefit from a collaborative approach including organizational scientists who have begun to examine team rewards in other contexts as discussed in Chapter 8 (e.g., Chen, Williamson, and Zhou, 2012) and experts in the science of team science.

Finally, we noted in Chapter 8 that there is a general lack of research on team science from an organizational perspective. Further research from this perspective would be valuable to inform research and practice. For example, the emergence of such new organizational forms as multi-team systems, cross-network scientific collaborations, and large, geographically dispersed research centers may require new approaches to team or group composition, professional development, and leadership. However, we noted in Chapter 6 that there has been little research to date on leadership in multi-team systems; only a few studies have begun to explore how system and team leaders can best foster coordination within and among the component science teams. Similarly, new organizational forms of team science are likely to present new challenges for composing and assembling the team or group, and for providing professional development.

FUNDING AND EVALUATION OF TEAM SCIENCE

We have noted that evaluating the processes and outcomes of team science is challenging, in part because science teams and larger groups may have multiple goals. Research is needed to develop new evaluative criteria that are appropriately matched to the respective goals and concerns of the teams, groups, organizations, institutions, funders, and community groups that have a stake in the foci, processes, and outcomes of the projects. In Chapter 9, we noted that federal scientific agencies are increasingly interested in examining their own processes, so that they can improve their practices and better address important social, technical, and scientific challenges. To date, however, very little empirical evidence is available from such efforts. Research is needed to help both public agencies and private foundations best deploy their resources to foster effective team science and find the optimal balance between team and non-team approaches. This research would provide answers to questions such as:

- How can funders and scientific review panels better identify team proposals that are likely to succeed or fail?
- What happens when the funding for a science team or group is withdrawn? Does the lack of long-term funding commitment lead researchers to revert to more traditional small, incremental scientific development processes? What is the relationship between the sustainability of funding and a supportive institutional context in terms of the likelihood of long-term success?
- What types of management, beyond the traditional funder roles of evaluating research proposals and requiring written reports, might facilitate science team effectiveness?
- Would team effectiveness be enhanced if funders provided ongoing technical assistance and emergency assistance to address collaboration challenges as they arise?

More specifically, research is needed to understand how alternative funding strategies may affect science team effectiveness. In Chapter 9, we recommended that funders require collaboration plans. Studies comparing the effectiveness of teams and groups that did and did not include a collaboration plan in their proposals would enable a learning and improvement-oriented approach to the management of team science.

Because peer review panels function as teams in and of themselves, research to better understand how their structure and dynamics influence reviews of team science proposals would provide useful information to funders. It would also be valuable to study how new approaches in which reviewers assemble "dream teams" with the goal of rapidly advancing science and translating discoveries in targeted areas affect the processes and outcomes of these teams.

SUMMARY, CONCLUSION, AND RECOMMENDATION

CONCLUSION. *Targeted research is needed to evaluate and refine the tools, interventions, and policies recommended above, along with more basic research, to guide continued improvement in the effectiveness of team science. However, few if any funding programs support research on the effectiveness of science teams and larger groups.*

RECOMMENDATION 9: **Public and private funders should support research on team science effectiveness through funding. As critical first steps, they should support ongoing evaluation and refinement of the interventions and policies recommended above and research on the role of scientific organizations (e.g., research centers, networks) in supporting science teams and larger groups. They should also collaborate with universities and the scientific community to facilitate researchers' access to key team science personnel and datasets.**

In closing, we note the promise of new research methods and approaches for advancing the research on team science effectiveness. In Chapter 2, we discussed the unique concerns of the science of team science, including its focus on highly diverse units of analysis, ranging from the individual to the team, the organization, and society as a whole and the need for developing valid, reliable metrics and criteria to understand and evaluate team processes and their relationships to scientific and translational outcomes. We noted that new research approaches and methods could help the field with these various concerns. For example, complexity theory offers a promising route to understand how behaviors, actions, and reactions at each level of a system affect actions at the other levels and the emergent behavior of the system as a whole. Researchers have begun to investigate team science using a complex adaptive system approach.

New methods are also available for studying team dynamics. For example, team or group members can be equipped with small electronic sensor badges (about the size of a smartphone) to record data on their interactions, including whether they are face-to-face, how close they are to one another, and the intensity of their conversation. Similarly, electronic communication data, such as emails and texts, can be recorded and analyzed. Data illuminating team or group dynamics—whether captured by unobtrusive sensors, through records of electronic communications, or through more traditional surveys—can be creatively combined with bibliometric data to examine the relationship between team processes and outcomes (in the form of scientific publications). Because team or group dynamics, goals, and outcomes change over time as science teams move through different phases in their work, longitudinal research designs coupled with analysis of temporally tagged data can provide greater insight than cross-sectional, one-time approaches.

Empirical research on science teams and groups can also benefit from simulation and modeling methods. Simulation allows technological tasks conducted by science teams and group in the real world (e.g., joint use of scientific equipment or virtual meeting technologies) to be studied under controlled laboratory conditions. In this way, technologies can be evaluated on the basis of their ability to improve science team effectiveness. Also, computation models (e.g., agent-based models, dynamical systems models, social network models) of findings regarding team member interactions under varying conditions in the literature on teams (including science teams) can help to extend empirical results from small science teams to larger groups and organizations.

References

Academy of Medical Sciences. (2013). *Team Science*. London: Author. Available: http://www. acmedsci.ac.uk/policy/policy-projects/team-science/ [April 2015].

Adams, J.D., Black, G.C., Clemmons, J.R., and Stephan, P.E. (2005). Scientific teams and institutional collaborations: Evidence from U.S. universities, 1981–1999. *Research Policy, 34*(3):259–285.

Adler, P.S., and Chen, C.X. (2011). Combining creativity and control: Understanding individual motivation in large-scale collaborative creativity. *Accounting, Organizations, and Society, 36*:63–85.

Alberts, B., Kirschner, M.W., Rilghman, S., and Varmus, H. (2014). Rescuing U.S. biomedical research from its systemic flaws: Perspective. *Proceedings of the National Academy of Sciences of the United States of America, 111*(16):5773–5777. Available: http://www. pnas.org/content/111/16/5773.full [May 2015].

Allen, L., Brand, A., Scott, J., Altman, M., and Hlava, M. (2014). Publishing: Credit where credit is due. *Nature, 508*(7496):312–313. Available: http://www.nature.com/news/ publishing-credit-where-credit-is-due-1.15033 [May 2015].

Allport, F.H. (1932). Psychology in relation to social and political problems. In P.S. Achilles (Ed.), *Psychology at Work* (pp. 199–252). New York: Whittlesey House. Available: https://www.brocku.ca/MeadProject/Allport/Allport_1932.html [May 2015].

Alonso, A., Baker, D.P., Holtzman, A., Day, R., King, H., Tommey, L., and Salas, E. (2006). Reducing medical error in the military health system: How can team training help? *Human Resource Management Review, 16*(3):396–415.

Altbach, P.G., Gumport, P.J., and Berdahl, R.O. (Eds). (2011) *American Higher Education in the Twenty-First Century: Social, Political, and Economic Challenges* (3rd ed.). Baltimore, MD: Johns Hopkins University Press.

Anacona, D., and Caldwell, D. (1992). Demography and design: Predictors of new product team performance. *Organizational Science, 3*:321–241.

Anderson, N., and West, M.A. (1998). Measuring climate for work group innovation: Development and validation of the team climate inventory. *Journal of Organizational Behavior, 19*:235–258.

Aranoff, D., and Bartkowiak, B. (2012). A review of the website TeamScience.net. *Clinical Medicine and Research*, 10(1):38–39. Available: http://www.clinmedres.org/content/10/1/38.full.pdf+html [October 2014].

Argyle, M., and Cook, M. (1976). *Gaze and Mutual Gaze*. New York: Cambridge University Press. Available: http://www.jstor.org/stable/27847843?seq=1 [October 2014].

Asencio, R., Carter, D.R., DeChurch, L.A., Zaccaro, S.J., and Fiore, S.M. (2012). Charting a course for collaboration: A multiteam perspective. *Translational and Behavioral Medicine*, 2(4):487–494. Available: http://www.ncbi.nlm.nih.gov/pmc/articles/PMC3717938/pdf/13142_2012_Article_170.pdf [October 2014].

Ashforth, B.E., and Mael, F. (1989). Social identity theory and the organization. *Academy of Management Review*, 14:20–39. Available: http://www.jstor.org/stable/pdfplus/258189.pdf?acceptTC=true&jpdConfirm=true [October 2014].

ATLAS Collaboration. (2012). Observation of a new particle in the search for the Standard Model Higgs boson with the ATLAS detector at the LHC. *Physics Letters B*, 716(1):1–29. Available: http://www.sciencedirect.com/science/article/pii/S037026931200857X [March 2014].

Austin, A.E. (2011). *Promoting Evidence-Based Change in Undergraduate Science Education*. Presented at the Fourth Committee Meeting on Status, Contributions, and Future Directions of Discipline-Based Education Research, National Research Council, Washington, DC. Available: http://sites.nationalacademies.org/DBASSE/BOSE/DBASSE_080124 [April 2015].

Austin, J.R. (2003). Transactive memory in organizational groups: The effects of content, consensus, specialization, and accuracy on group performance. *Journal of Applied Psychology*, 88(5):866–878.

Avolio, B.J., Reichard, R.J., Hannah, S.T., Walumbwa, F.O., and Chan, A. (2009). A meta-analytic review of leadership impact research: Experimental and quasi-experimental studies. *The Leadership Quarterly*, 20:764–784.

Bacharach, S.B., Bamberger, P., and Mundell, B. (1993). Status inconsistency in organizations: From social hierarchy to stress. *Journal of Organizational Behavior*, 14:21–36. Available: http://arno.uvt.nl/show.cgi?fid=113234 [October 2014].

Bain, P.G., Mann, L., and Pirola-Merlo, A. (2001). The innovation imperative: The relationship between team climate, innovation, and performance in research and development teams. *Small Group Research*, 32(1):55–73. Available: http://sgr.sagepub.com/content/32/1/55.full.pdf [October 2014].

Bainbridge, W.S. (2007). The scientific research potential of virtual worlds. *Science*, 317(5837): 472–476. Available: http://www.sciencemag.org/content/317/5837/472.full.pdf [October 2014].

Baker D.P., Day, R., and Salas, E. (2006). Teamwork as an essential component of high-reliability organizations. *Health Services Research*, 41(4 Pt 2):1576–1598.

Bandura, A. (1977). Self-efficacy: Toward a unifying theory of behavioral change. *Psychological Review*, 84(2):191–215. Available: http://www.uky.edu/~eushe2/Bandura/Bandura1977PR.pdf [October 2014].

Bass, B.M. (1985). *Leadership and Performance Beyond Expectations*. New York: Free Press.

Bass, B.M., and Riggio, R.E. (2006). *Transformational Leadership* (2nd ed.). Mahwah, NJ: Lawrence Erlbaum.

Beal, D.J., Cohen, R.R., Burke, M.J., and McLendon, C.L. (2003). Cohesion and performance in groups: A meta-analytic clarification of construct relations. *Journal of Applied Psychology*, 88(6):989–1004.

Bear, J.B., and Woolley, A.W. (2011). The role of gender in team collaboration and performance. *Interdisciplinary Science Reviews*, 36(2):146–153.

Becker, F. (2004). *Offices at Work: Uncommon Workspace Strategies That Add Value and Improve Performance*. San Francisco: Jossey-Bass.

Bell, B.S., and Kozlowski, S.W.J. (2002). A typology of virtual teams: Implications for effective leadership. *Group and Organization Management*, 27(1):14–19.

Bell, B.S., and Kozlowski, S.W.J. (2011). Collective failure: The emergence, consequences, and management of errors in teams. In D.A. Hoffman and M. Frese (Eds.), *Errors in Organizations* (pp. 113–141). New York: Routledge.

Bell, E., Canuto, M., and Sharer, R. (Eds). (2003). *Understanding Early Classic Copan*. Philadelphia: University of Pennsylvania Museum Press.

Bell, S.T., Villago, A.J., Lukasik, M.A., Belau, L., and Briggs, A.L. (2011). Getting specific about demographic diversity variable and team performance relationships: A meta-analysis. *Journal of Management*, 37(3):709–743. Available: http://jom.sagepub.com/content/37/3/709.full.pdf [October 2014].

Bennett, L.M., and Gadlin, H. (2012). Collaboration and team science: From theory to practice. *Journal of Investigative Medicine*, 60(5):768–775.

Bennett. L.M., and Gadlin, H. (2014). Supporting interdisciplinary collaboration: The role of the institution. In M. O'Rourke, S. Crowley, S.D. Eigenbrode, and J.D. Wulfhorst (Eds.), *Enhancing Communication and Collaboration in Interdisciplinary Research* (pp. 356-384). Thousand Oaks, CA: Sage.

Bennett, L.M., Gadlin, H., and Levine-Finley, S. (2010). *Collaboration and Team Science: A Field Guide*. NIH Publication No. 10-7660. Bethesda, MD: National Institutes of Health. Available: http://ombudsman.nih.gov/collaborationTS.html [May 2015].

Berrett, D. (2011). Tenure across borders. *Inside Higher Education*. Available: http://www.insidehighered.com/news/2011/07/22/usc_rewards_collaborative_and_interdisciplinary_work_among_faculty#sthash.LPOPUxoq.dpbs [May 2015].

Beyer, H., and Holtzblatt, K. (1998). *Contextual Design: Defining Customer-Centered Systems*. San Francisco: Morgan Kaufmann.

Bezrukova, K. (2013). *Understanding and Addressing Faultlines*. Presented at the National Research Council's Workshop on Science Team Dynamics and Effectiveness, July 1, Washington, DC. Available: http://sites.nationalacademies.org/cs/groups/dbassesite/documents/webpage/dbasse_083763.pdf [May 2015].

Bezrukova, K., Jehn, K.A., Zanutto, E.L., and Thatcher, S.M.B. (2009). Do workgroup faultlines help or hurt? A moderated model of faultlines, team identification, and group performance. *Organization Science*, 20(1):35–50.

Bezrukova, K., Thatcher, S.M.B., Jehn, K., and Spell, C. (2012). The effects of alignments: Examining group faultiness, organizational cultures, and performance. *Journal of Applied Psychology*, 97(1):77–92.

Binford, L. (1978). Dimensional analysis of behavior and site structure: Learning from an Eskimo hunting stand. *American Antiquity*, 43(3):330–361. Available: http://www.jstor.org/stable/pdfplus/279390.pdf?&acceptTC=true&jpdConfirm=true [October 2014].

Binford, L. (1980). Willow smoke and dogs' tails: Hunter-gatherer settlement systems and archaeological site formation. *American Antiquity*, 45(1):4–20. Available: http://www.jstor.org/stable/pdfplus/279653.pdf?&acceptTC=true&jpdConfirm=true [October 2014].

Binford, L. (2001). *Constructing Frames of Reference: An Analytical Method for Archaeological Theory Building Using Ethnographic and Environmental Data Sets*. Berkeley and Los Angeles: University of California Press.

Blackburn, R., Furst, S.A., and Rosen, B. (2003). Building a winning virtual team: KSA's, selections, training, and evaluation. In C.B. Gibson and S.G. Cohen (Eds.), *Virtual Teams That Work: Creating Conditions for Virtual Team Effectiveness*. San Francisco: Jossey-Bass.

Blickensderfer, E., Cannon-Bowers, J.A., and Salas, E. (1997). Theoretical bases for team self-corrections: Fostering shared mental models. In M.M. Beyerlein and D.A. Johnson (Eds.), *Advances in Interdisciplinary Studies of Work Teams* (vol. 4, pp. 249–279). Greenwich, CT: JAI Press.

Bonney, R., Cooper, C.B., Dickinson, J., Kelling, S., Phillips, T., Rosernberg, K.V., and Shirk, J. (2009). Citizen science: A developing tool for expanding science knowledge and scientific literacy. *Bioscience, 59*(11):977–984.

Borgman, C.L. (2015). *Big Data, Little Data, No Data: Scholarship in the Networked World.* Cambridge, MA: MIT Press.

Börner, K., Contractor, N., Falk-Krzesinski, H.J., Fiore, S.M., Hall, K.L., Keyton, J., et al. (2010). A multi-level systems perspective for the science of team science. *Science Translational Medicine, 2*(49):1–5. Available: http://cns.iu.edu/images/pub/2010-borner-et-al-multi-level-teamsci.pdf [October 2014].

Börner, K., Conlon, M., Coron-Rikert, J., and Ding, Y. (2012). *VIVO: A Semantic Approach to Scholarly Networking and Discovery.* San Rafael, CA: Morgan & Claypool. Available: http://www.morganclaypool.com/doi/abs/10.2200/S00428ED1V01Y201207WBE002 [May 2014].

Borrego, M., and Newsander, L.K. (2010). Definitions of interdisciplinary research: Toward graduate-level interdisciplinary learning outcomes. *The Review of Higher Education, 34*(1):61–84.

Borrego, M., Karlin, J., McNair, L.D., and Beddoes, K. (2013). Team effectiveness theory from industrial and organizational psychology applied to engineering student project teams—A review. *Journal of Engineering Education, 102*(4):472–512.

Borrego, M., Boden, D., Pietrocola, D., Stoel, C.F., Boone, R.D., and Ramasubramanian, M.K. (2014). Institutionalizing interdisciplinary graduate education. In M. O'Rourke, S. Crowley, S.D. Eigenbrode, and J.D. Wulfhorst (Eds.), *Enhancing Communication and Collaboration in Interdisciplinary Research* (pp. 335–355). Thousand Oaks, CA: Sage.

Bos, N. (2008). Motivation to contribute to collaboratories: A public goods approach. In G.M. Olson, A. Zimmerman, and N. Bos (Eds.), *Scientific Collaboration on the Internet* (pp. 251–274). Cambridge, MA: MIT Press. Available: http://cryptome.org/2013/01/aaron-swartz/Internet-Scientific-Collaboration.pdf [October 2014].

Bos, N., Olson, G.M., and Zimmerman, A. (2008). Final thoughts: Is there a science of collaboratories? In G.M. Olson, A. Zimmerman, and N. Bos (Eds.), *Scientific Collaboration on the Internet* (pp. 377–393). Cambridge, MA: MIT Press. Available: http://cryptome.org/2013/01/aaron-swartz/Internet-Scientific-Collaboration.pdf [October 2014].

Boudreau, K., Brady, T., Gangulia, I., Gaule, P., Guinan, E., Hollenberg, A., and Lakhani, K. (2012*). Co-location and Scientific Collaboration: Evidence from a Field Experiment.* Harvard University Working Paper, Cambridge, MA.

Boudreau, K.J., Guinan, E., Lakhani, K.R., and Riedl, C. (2014). Looking across and looking beyond the knowledge frontier: Intellectual distance and resource allocation in science. *Management Science,* forthcoming. Available: http://ssrn.com/abstract=2478627 [May 2015].

Bowers, C.A., Jentsch, F., and Salas, E. (2000). Establishing aircrew competencies: A comprehensive approach for identifying CRM training needs. In H.F. O'Neil and D. Andrews (Eds.), *Aircrew Training and Assessment* (pp. 67–84). Mahwah, NJ: Lawrence Erlbaum.

Boyce, L., Zaccaro, S.J., and Wisecarver, M. (2010). Propensity for self-development of leadership attributes: Understanding, predicting, and supporting performance of leader self-development. *The Leadership Quarterly, 21*(1), 159–178.

Bozeman, B., and Boardman, C. (2013). *An Evidence-Based Assessment of Research Collaboration and Team Science: Patterns in Industry and University-Industry Partnerships.* Presented at the National Research Council's Workshop on Institutional and Organizational Supports for Team Science, October 24, Washington, DC. Available: http://sites. nationalacademies.org/DBASSE/BBCSS/DBASSE_085236 [May 2015].

Bozeman, B., Fay, D., and Slade, C.P. (2012). Research collaboration in universities and academic entrepreneurship: The state-of-the-art. *Journal of Technological Transfer, 38*(1): 1–67.

Bozeman, B., and Gaughan, M. (2011). How do men and women differ in research collaborations? An analysis of the collaboration motives and strategies of academic researchers. *Research Policy, 40*(10):1393–1402.

Brannick, M.T., Prince, A., Prince, C., and Salas, E. (1995). The measurement of team process. *Human Factors, 37*(3):641–651. Available: http://hfs.sagepub.com/content/37/3/641.full. pdf [October 2014].

Braun, D. (1998). The role of funding agencies in the cognitive development of science. *Research Policy, 27*:807–821.

Brewer, M.B. (1999). The psychology of prejudice: Ingroup love or outgroup hate? *Journal of Social Issues, 55*(3):429–444. Available: http://onlinelibrary.wiley.com/doi/10.1111/0022-4537.00126/pdf [October 2014]

Brill, M., Weidemann, S., and BOSTI Associates. (2001). *Disproving Widespread Myths about Workplace Design.* Jasper, IN: Kimball International.

Broad, W.J. (2014). Billionaires with big ideas are privatizing American science. *The New York Times,* March 14. Available: http://www.nytimes.com/2014/03/16/science/billionaires-with-big-ideas-are-privatizing-american-science.html?_r=1 [May 2014].

Brown, J.S., and Thomas, D. (2006). You play World of Warcraft? You're hired. *Wired, 14*(4). Available: http://archive.wired.com/wired/archive/14.04/learn.html [October 2014].

Burke, C.S., Stagl, K.C., Klein, C., Goodwin, G.F., Salas, E., and Halpin, S.M. (2006). What type of leadership behaviors are functional in teams? A meta-analysis. *Leadership Quarterly, 17*:288–307.

Burns, T., and Stalker, G.M. (1966). *The Management of Innovation* (2nd ed.). London: Tavistock.

Cameron, A.F., and Webster, J. (2005). Unintended consequences of emerging communication technologies: Instant messaging in the workplace. *Computers in Human Behavior, 21*(1):85–103.

Campion, M.A., Medsker, G.J., and Higgs, A.C. (1993). Relations between work-group characteristics and effectiveness: Implications for designing effective work groups. *Personnel Psychology, 46*(4):823–847. Available: http://www.krannert.purdue.edu/faculty/campionm/Relations_Between_Work.pdf [October 2014]

Cannon-Bowers, J.A. (2007). Fostering mental model convergence through training. In F. Dansereau and F. Yammarino (Eds.), *Multi-Level Issues in Organizations and Time* (pp. 149–157). San Diego, CA: Elsevier.

Cannon-Bowers, J.A., and Salas, E. (1998). *Making Decisions Under Stress: Implications for Individual and Team Training.* Washington, DC: American Psychological Association.

Cannon-Bowers, J.A., Salas, E., and Converse, S.A. (1993). Shared mental models in expert team decision making. In N.J. Castellan (Ed.), *Individual and Group Decision Making: Current Issues* (pp. 221–246). Hillsdale, NJ: Lawrence Erlbaum.

Cannon-Bowers, J.A., Tannenbaum, S.I., Salas, E., and Volpe, C.E. (1995). Defining team competencies and establishing team training requirements. In R. Guzzo and E. Salas (Eds.), *Team Effectiveness and Decision Making in Organizations* (pp. 333–380). San Francisco: Jossey-Bass.

Cannon-Bowers, J.A., Salas, E., Blickensderfer, E., and Bowers, C.A. (1998). The impact of cross-training and workload on team functioning: A replication and extension of initial findings. *Human Factors, 40*(1):92–101. Available: http://hfs.sagepub.com/content/40/1/92.full.pdf [October 2014].

Carley, K., and Wendt, K. (1991). Electronic mail and scientific communication: A study of the Soar extended research goup. *Knowledge: Creation, Diffusion, Utilization, 12*(4):406–440. Available: http://scx.sagepub.com/content/12/4/406.full.pdf [October 2014].

Carney, J., and Neishi, K. (2010). *Bridging Disciplinary Divides: Developing an Interdisciplinary STEM Workforce.* Prepared for Division of Graduate Education, National Science Foundation by Abt Associates. Available: http://www.abtassociates.com/reports/ES_IGERT_SUMMARY_REPORT_October_2010.pdf [May 2014].

Carr, J.Z., Schmidt, A.M., Ford, J.K., and DeShon, R.P. (2003). Climate perceptions matter: A meta-analytic path analysis relating molar climate, cognitive and affective states, and individual level work outcomes. *Journal of Applied Psychology, 88*(4):605–619.

Carroll, J.B., Rosson, M.B., and Zhou, J. (2005). Collective efficacy as a measure of community. In *CHI 2005 Proceedings of the SIGCHI Conference on Human Factors in Computing Systems* (pp. 1–10). Available: http://dl.acm.org/citation.cfm?id=1054974 [June 2014].

Carton, A.M., and Cummings, J.N. (2012). A theory of subgroups in work teams. *Academy of Management Review, 37*(3):441–470.

Carton, A.M., and Cummings, J.N. (2013). The impact of subgroup type and subgroup configurational properties on work team performance. *Journal of Applied Psychology, 98*(5):732–758.

Case Western University School of Medicine (2014). *Draft Team Science Promotion and Tenure Process.* Available: http://casemed.case.edu/ctsc/teamscience/ [May 2014].

Cash, D.W., Clark, W.C., Alcock, F., Dickson, M.N., Eckley, N., Guston, D.H., Jäger, J., and Mitchell, R.B. (2003). Knowledge systems for sustainable development. *Proceedings of the National Academy of Sciences of the United States of America, 100*(14):8086–8091.

Catalini, C. (2012). *Microgeography and the Direction of Inventive Activity.* Rotman School of Management Working Paper No. 2126890, University of Toronto. Available SSRN: http://ssrn.com/abstract=2126890 or http://dx.doi.org/10.2139/ssrn.2126890 [June 2015].

Chao, G.T., and Moon, H. (2005). The cultural mosaic: A meta-theory for understanding the complexity of culture. *Journal of Applied Psychology, 90*(6):1128–1140.

Chen, C.S., Williamson, M.G., and Zhou, F.H. (2012). Reward system design and group creativity: An experimental investigation. *Accounting Review, 87*(6):1885–1911.

Chen, G., and Bliese, P.D. (2002). The role of different levels of leadership in predicting self- and collective efficacy: Evidence for discontinuity. *Journal of Applied Psychology, 87*(3):549–556.

Chen, G. and Tesluk, P. (2012). Team participation and empowerment: A multilevel perspective. In S.W.J. Kozlowski (Ed.), *Oxford Handbook of Organizational Psychology* (vol. 2). Cheltenham, UK: Oxford University Press.

Chen, G., Thomas, B., and Wallace, J.C. (2005). A multilevel examination of the relationships among training outcomes, mediating regulatory processes, and adaptive performance. *Journal of Applied Psychology, 90*(5):827–841.

Chen, G., Kanfer, R., DeShon, R.P., Mathieu, J.E., and Kozlowski, S.W.J. (2009). The motivating potential of teams: Test and extension of Chen and Kanfer's (2006) cross-level model of motivation in teams. *Organizational Behavior and Human Decision Processes, 110*:45–55. Available: http://scs.math.yorku.ca/images/4/40/ICC_for_aggregation.pdf [October 2014].

Chen, G., Farh, J.-L., Campbell-Bush, E.M., Wu, Z., and Wu, X. (2013). Teams as innovative systems: Multilevel motivational antecedents of innovation in R&D teams. *Journal of Applied Psychology, 98*(6):1018–1027.

Chen, Y., and Sonmez, T. (2006). School choice: An experimental study. *Journal of Economic Theory,* 127:202–231. Available: http://yanchen.people.si.umich.edu/papers/school_choice_jet2006.pdf [October 2014].

Chompalov, I., Genuth, J., and Shrum, W. (2002) The organization of scientific collaborations. *Research Policy,* 31(5):749–767.

Chubin, D.E., Derrick, E., Feller, I., and Phartiyal, P. (2009). *AAAS Review of the NSF Science and Technology Centers Integrative Partnerships (STC) Program, 2000–2009.* Washington, DC: American Association for the Advancement of Science. Available: http://www.aaas.org/report/final-report-aaas-review-nsf-science-and-technology-centers-integrative-partnerships-stc [May 2014].

Claggett, J., and Berente, N. (2012). *Organizing for Digital Infrastructure Innovation: The Interplay of Initiated and Sustained Attention.* Presented at the Hawaiian International Conference on System Sciences (HICSS-45), Maui, January 4–7. Available: http://ieeexplore.ieee.org/stamp/stamp.jsp?tp=&arnumber=6149529 [October 2014].

CMS Collaboration (2012). Observation of a new boson at a mass of 125 GeV with the CMS experiment at the LHC. *Physics Letters B,* 716(1):30-61. Available: http://www.sciencedirect.com/science/article/pii/S0370269312008581 [March 2014].

COALESCE. (2010). *CTSA Online Assistance for Leveraging the Science of Collaborative Effort.* Department of Preventive Medicine, Feinberg School of Medicine, Northwestern University. Available: http://www.teamscience.net/ [May 2014].

Cohen, J. (1992). A power primer. *Psychological Bulletin, 112*(1):155–159.

Collins, D.B., and Holton, E.F. (2004). The effectiveness of managerial leadership development programs: A meta-analysis of studies from 1982 to 2001. *Human Resource Development Quarterly,* 15:217–248.

Collins, R. (1998). *The Sociology of Philosophies: A Global Theory of Intellectual Change.* Cambridge, MA: The Belknap Press of Harvard University Press.

Colorado Clinical and Translational Sciences Institute. (2014). *Leadership for Innovative Team Science (LITeS): Description and Directory 2014–2015.* Available: http://www.ucdenver.edu/research/CCTSI/education-training/LITeS/Documents/LITeS2014-2015Directory.pdf [December 2014].

Contractor, N.S. (2013). *Some Assembly Required: Organizing in the 21st Century.* Presented at the National Research Council's Workshop on Science Team Dynamics and Effectiveness, July 1, Washington, DC. Available: http://sites.nationalacademies.org/DBASSE/BBCSS/DBASSE_083679 [September 2014].

Contractor, N.S., DeChurch, L.A., Asencio, R., Huang, Y., Murase, T., and Sawant, A. (2014). *Enabling Teams to Self-Assemble: The MyDreamTeam Builder.* Presented in symposium titled Enhancing Team Effectiveness Across and Between Levels of Analysis (J. Methot and J.E. Mathieu, co-chairs) at the Society for Industrial and Organizational Psychology Annual Meeting, May 15, Honolulu, HI.

Cooke, N.J., and Gorman, J.C. (2009). Interaction-based measures of cognitive systems. *Journal of Cognitive Engineering and Decision Making: Special Section on Integrating Cognitive Engineering in the Systems Engineering Process: Opportunities, Challenges and Emerging Approaches* 3:27–46. Available: http://edm.sagepub.com/content/3/1/27.full.pdf+html [October 2014].

Cooke, N.J., Salas, E., Cannon-Bowers, J.A., and Stout, R. (2000). Measuring team knowledge. *Human Factors,* 42:151–173.

Cooke, N.J., Kiekel, P.A., and Helm, E. (2001). Measuring team knowledge during skill acquisition of a complex task. *International Journal of Cognitive Ergonomics,* 5(3):297–315.

Cooke, N.J., Gorman, J.C., and Kiekel, P.A. (2008). Communication as team-level cognitive processing. In M. Letsky, N. Warner, S. Fiore, and C. Smith (Eds.), *Macrocognition in Teams: Theories and Methodologies* (pp. 51–64). Hants, UK: Ashgate.

Cooke, N.J., Gorman, J.C., Myers, C.W., and Duran, J.L. (2013). Interactive team cognition. *Cognitive Science, 37(2):255–285.*

Cramton, C. (2001). The mutual knowledge problem and its consequences for dispersed collaboration. *Organization Science, 12(3):346–371.*

CRediT. (2015). *CRediT: An Open Standard for Expressing Roles Intrinsic to Research.* London: Author. Available: http://credit.casrai.org/ [April 2015].

Cronin, M.A., Weingart, L.R., and Todorova, G. (2011). Dynamics in groups: Are we there yet? *Academy of Management Annals, 5(1):571–612.*

Crow, M.M. (2010). Organizing teaching and research to address the grand challenges of sustainable development. *BioScience, 60(7):488–489.* Available: https://president.asu.edu/sites/default/files/BioScience%20Article%20070110%20Organizing%20to%20Address%20Grand%20Challenges%20Sustainable%20Dev.pdf [October 2014].

Crow, M.M., and Debars, W.B. (2014). Interdisciplinarity as a design problem: Toward mutual intelligibility among academic disciplines in the American research university. In M. O'Rourke, S. Crowley, S.D. Eigenbrode, and J.D. Wulfhorst (Eds.), *Enhancing Communication and Collaboration in Interdisciplinary Research.* Thousand Oaks, CA: Sage.

Crowston, K. (2013). *Response to a Technology Framework to Support Team Science.* Presented at the National Research Council's Workshop on Institutional and Organization Supports for Team Science, October 24, Washington, DC. Available: http://sites.nationalacademies.org/DBASSE/BBCSS/DBASSE_085236l [March 2014].

Croyle, R.T. (2008). The National Cancer Institute's Transdisciplinary Centers Initiative and the need for building a science of team science. *American Journal of Preventive Medicine, 35(2S):S90–S93.* Available: http://www.sciencedirect.com/science/article/pii/S0749379708004224 [October 2014].

Croyle, R.T. (2012). Confessions of a team science funder. *Translational Behavioral Medicine, 2(4):531–534.* Available: http://www.ncbi.nlm.nih.gov/pmc/articles/PMC3717939/ [October 2014].

Csikszentmihalyi, M. (1994). The domain of creativity. In D.H. Feldman, M. Csikszentmihalyi, and H. Gardner (Eds.), *Changing the World: A Framework for the Study of Creativity* (pp. 138–158). London: Praeger.

Cummings, J.N., and Haas, M.R. (2012). So many teams, so little time: Time allocation matters in geographically dispersed teams. *Journal of Organizational Behavior, 33(3):316–341.*

Cummings, J.N., and Kiesler, S. (2005). Collaborative research across disciplinary and institutional boundaries. *Social Studies of Science, 35(5):703–722.* Available: http://sss.sagepub.com/content/35/5/703.full.pdf [October 2014].

Cummings, J.N., and Kiesler, S. (2007). Coordination costs and project outcomes in multi-university collaborations. *Research Policy, 36(10):1620–1634.*

Cummings, J.N., and Kiesler, S. (2008). Who collaborates successfully? Prior experience reduces collaboration barriers in distributed interdisciplinary research. In *Proceedings of the 2008 Conference on Computer-Supported Collaborative Work* (pp. 437–446). New York: ACM. Available: http://citeseerx.ist.psu.edu/viewdoc/download?doi=10.1.1.352.4322&rep=rep1&type=pdf [May 2014].

Cummings, J.N., and Kiesler, S. (2011) *Organization Theory and New Ways of Working in Science.* Presented at the Atlanta Conference on Science and Innovation Policy, September 15–17. Available: http://ieeexplore.ieee.org/xpl/articleDetails.jsp?arnumber=6064466 [May 2015].

Cummings, J.N., Espinosa, J.A., and Pickering, C.K. (2009). Crossing spatial and temporal boundaries in globally distributed projects: A relational model of coordination delay. *Information Systems Research, 30*(3):420–439.

Cummings, J.N., Kiesler, S., Zadeh, R., and Balakrishnan, A. (2013). Group heterogeneity increases the risks of large group size: A longitudinal study of productivity in research groups. *Psychological Science, 24*(6):880–890.

Day, D.V. (2010). The difficulties of learning from experience and the need for deliberate practice. *Industrial and Organizational Psychology, 3*(1):41–44.

Day, D.V. (2011). Integrative perspectives on longitudinal investigations of leader development: From childhood through adulthood. *The Leadership Quarterly, 22*(3):561–571.

Day, D.V., and Antonakis, J. (2012). Leadership: Past, present, and future. In D.V. Day and J. Antonakis (Eds.), *The Nature of Leadership* (2nd ed., pp. 3–25). Thousand Oaks, CA: Sage.

Day, D.V., and Harrison, M.M. (2007). A multilevel, identity-based approach to leadership development. *Human Resource Management Review, 17*(4):360–373.

Day, D.V., and Sin, H-P. (2011). Longitudinal tests of an integrative model of leader development: Charting and understanding developmental trajectories. *The Leadership Quarterly, 22*(3):545–560.

Day, D.V., and Zaccaro, S.J. (2007). Leadership: A critical historical analysis of the influence of leader traits. In L.L. Koppes (Ed.), *Historical Perspectives in Industrial and Organizational Psychology* (pp. 383–405). Mahwah, NJ: Lawrence Erlbaum.

Day, D.V., Gronn, P., and Salas, E. (2004). Leadership capacity in teams. *Leadership Quarterly, 15*:857–880. Available: http://www.themedfomscu.org/media/Leadership_Capacity.pdf [October 2014].

Day, D.V., Sin, H-P., and Chen, T.T. (2004). Assessing the burdens of leadership: Effects of formal leadership roles on individual performance over time. *Personnel Psychology, 57*(3):573–605. Available: http://onlinelibrary.wiley.com/doi/10.1111/j.1744-6570.2004.00001.x/pdf [October 2014].

DeChurch, L.A., and Marks, M.A. (2006). Leadership in multiteam systems. *Journal of Applied Psychology, 91*(2):311–329. Available: http://www.delta.gatech.edu/papers/leadership Multiteam.pdf [October 2014].

DeChurch, L.A., and Mesmer-Magnus, J.R. (2010). The cognitive underpinnings of effective teamwork: A meta-analysis. *Journal of Applied Psychology, 95*(1):32–53. Available: http://www.owlnet.rice.edu/~ajv2/courses/12c_psyc438001/DeChurch,%20&%20 Mesmer-Magnus%20(2010)%20JAP.pdf [October 2014].

DeChurch, L.A., and Zaccaro, S.J. (2013). *Innovation in Scientific Multiteam Systems: Confluent and Countervailing Forces.* Presented at the National Research Council's Workshop on Science Team Dynamics and Effectiveness, Washington, DC. Available: http://sites. nationalacademies.org/DBASSE/BBCSS/DBASSE_083679 [May 2014].

DeChurch, L.A., Burke, C.S., Shuffler, M.L., Lyons, R., Doty, D., and Salas, E. (2011). A historiometric analysis of leadership in mission critical multiteam environments. *Leadership Quarterly, 22*(1):152–169.

De Dreu, C.K.W., and Weingart, L.R. (2003). Task versus relationship conflict, team performance, and team member satisfaction: A meta-analysis. *Journal of Applied Psychology, 88*(4):741–749. Available: http://web.mit.edu/curhan/www/docs/Articles/15341_ Readings/Negotiation_and_Conflict_Management/De_Dreu_Weingart_Task-conflict_ Meta-analysis.pdf [October 2014].

Defila, R., DiGiulio, A., and Scheuermann, M. (2006). Forschungsverbundmanagement. *Handbuch for die Gestaltung inter- und transdisziplinärer Projeckte.* Zürich: vdr Hochschulverlag an der ETH Zürich. Cited in Stokols, D., Hall, K.L., Moser, R.P., Feng, A., Misra, S., and Taylor, B.K. (2010). Cross-disciplinary team science initiatives: Research, training, and translation. In R. Frodeman, J.T. Klein, C. Mitcham, and J.B. Holbrook (Eds.), *The Oxford Handbook of Interdisciplinarity.* Oxford, UK: Oxford University Press.

Delise, L.A., Gorman, C.A., and Brooks, A.M. (2010). The effects of team training on team outcomes: A meta-analysis. *Performance Improvement Quarterly, 22*(4):53–80. Available: http://onlinelibrary.wiley.com/doi/10.1002/piq.20068/pdf [October 2014].

DeRue, D.S. (2011). Adaptive leadership theory: Leading and following as a complex adaptive process. *Research in Organizational Behavior, 3*:125–150.

DeShon, R.P., Kozlowski, S.W.J., Schmidt, A.M., Milner, K.R., and Wiechmann, D. (2004). A multiple goal, multilevel model of feedback effects on the regulation of individual and team performance. *Journal of Applied Psychology, 89*(6):1035–1056.

Devine, D.J., and Philips, J.L. (2001). Do smarter teams do better? A meta-analysis of cognitive ability and team performance. *Small Group Research, 32*:507–532. Available: http://sgr.sagepub.com/content/32/5/507.full.pdf [October 2014].

deWit, F.R., Greer, L.L., and Jehn, K.A. (2012). The paradox of intergroup conflict: A meta-analysis. *Journal of Applied Psychology, 97*(2):360–390.

Dickinson, J.L., and Bonney, R. (2012). *Citizen Science: Public Participation in Environmental Research.* Ithaca, NY: Comstock.

Djorgovski, S.G., Hut, P., McMillan, S., Vesperini, E., Knop, R., Farr, W., and Graham, M.J. (2010). Exploring the use of virtual worlds as a scientific research platform: The Meta-Institute for Computational Astrophysics (MICA). In F. Lehmann-Grube, J. Sablatnig, O. Akan, P. Bellavista, J. Cao, F. Dressler, and D. Ferrari (Eds.), *Facets of Virtual Environments* (vol. 33, pp. 29–43). Berlin: Springer.

Doan, A., Ramakrishnan, R., and Halevy, A.Y. (2011). Crowdsourcing systems on the World-Wide Web. *Communications of the ACM, 4*(4):86–96.

Doherty-Sneedon, G., Anderson, A., O'Malley, C., Langton, S., Garrod, S., and Bruce, V. (1997). Face-to-face and video mediated communication: A comparison of dialogue structure and task performance. *Journal of Experimental Psychology: Applied, 3*(2):105–123.

Doorley, S., and Witthoft, S. (2012). *Make Space: How to Set the Stage for Creative Collaboration.* Hoboken, NJ: John Wiley & Sons.

Drath, W.H., McCauley, C.D., Palus, C.J., Van Velsor, E., O'Connor, P.M.G., and McGuire, J.B. (2008). Direction, alignment, commitment: Toward a more integrative ontology of leadership. *Leadership Quarterly, 19*(6):635–653.

Druss, B.G., and Marcus, S.C. (2005). Tracking publication outcomes of NIH grants. *American Journal of Medicine, 118*(6):658–663.

Duarte, D., and Snyder, N. (1999). *Mastering Virtual Teams.* San Francisco: Jossey-Bass.

Duderstadt, J.J. (2000). *A University for the Twenty-First Century.* Ann Arbor: University of Michigan Press.

Dust, S. B., and Zeigert, J. C. (2012). When and how are multiple leaders most effective? It's complex. *Industrial and Organizational Psychology: Perspectives on Science and Practice, 5*(4):421–424.

Dweck, C.S. (1986). Motivational processes affecting learning. *American Psychologist, 41*(10):1040–1048. Available: http://www.nisdtx.org/cms/lib/TX21000351/Centricity/Domain/21/j%20carlisle/Motivational%20Processes.pdf [October 2014].

Dyer, W.G., Dyer, W.G. Jr., and Dyer, J.H. (2007). *Team Building Proven Strategies for Improving Team Performance* (4th ed.). San Francisco: Jossey-Bass.

Edmondson, A.C. (1996). Learning from mistakes is easier said than done: Group and organizational influences on the detection and correction of human error. *Journal of Applied Behavioral Science, 32*:5–28.

Edmondson, A.C. (1999). Psychological safety and learning behavior in work teams. *Administrative Science Quarterly, 44*:350–383. Available: http://www.iacmr.org/Conferences/WS2011/Submission_XM/Participant/Readings/Lecture9B_Jing/Edmondson,%20ASQ%201999.pdf [October 2014].

Edmondson, A.C. (2002). The local and variegated nature of learning in organizations: A group-level perspective. *Organization Science, 13*(2):128–146.

Edmondson, A.C. (2003). Speaking up in the operating room: How team leaders promote learning in interdisciplinary action teams. *Journal of Management Studies, 40*(6):1419–1452. Available: http://onlinelibrary.wiley.com/doi/10.1111/1467-6486.00386/pdf [October 2014].

Edmondson, A.C., and Nembhard, I. (2009). Product development and learning in project teams: The challenges are the benefits. *Journal of Production Innovation Management, 26*(2): 123–138. Available: http://onlinelibrary.wiley.com/doi/10.1111/j.1540-5885.2009.00341.x/pdf [October 2014].

Edmondson, A.C., Bohmer, R.M., and Pisano, G.P. (2001). Disrupted routines: Team learning and new technology implementation in hospitals. *Administrative Science Quarterly, 46*(4):685–716. Available: http://web.mit.edu/curhan/www/docs/Articles/15341_Readings/Organizational_Learning_and_Change/Edmondson_et_al_Disrupted_Routines.pdf [October 2014].

Edmondson, A.C., Dillon, J.R., and Roloff, K.S. (2008). Three perspectives on team learning. In A. Brief and J. Walsh (Eds.), *The Academy of Management Annals, 1*(1), 269–314. Available: http://www.hbs.edu/faculty/Publication%20Files/07-029.pdf [May 2015].

Eigenbrode, S.D., O'Rourke, M., Wulfhorst, J.D., Althoff, D. M., Goldberg, C.S., Merrill, K., Morse, W., Nielsen-Pincus, M., Stephens, J., Winowiecki, L., and Bosque-Perez, N.A. (2007). Employing philosophical dialogue in collaborative science. *BioScience, 57*(1):55–64.

Ekmekci, O., Lotrecchiano, G.R., and Corcoran, M. (2014). The devil is in the (mis)alignment: Developing curriculum for clinical and translational science professionals. *Journal of Translational Medicine & Epidemiology, 2*(2):1029.

Ellis, A.P.J. (2006). System breakdown: The role of mental models and transactive memory in the relationship between acute stress and team performance. *Academy of Management Journal, 49*(3):576–589.

Engel, D., Woolley, A.W., Jing, L.S., Chabris, C.F., and Malone, T.W. (2014). Reading the mind in the eyes or reading between the lines? Theory of mind predicts collective intelligence equally well online and face-to-face. *PLOS One*. December 16. Available: http://journals.plos.org/plosone/article?id=10.1371/journal.pone.0115212 [January 2015].

Ensley, M.D., Hmielski, K.M., and Pearce, C.L. (2006). The importance of vertical and shared leadership within new venture top management teams: Implications for the performance of startups. *Leadership Quarterly, 17*(3):217–231. Available: http://digitalcommons.unl.edu/cgi/viewcontent.cgi?article=1073&context=managementfacpub [October 2014].

Entin, E.E., and Serfaty, D. (1999). Adaptive team coordination. *Human Factors: The Journal of the Human Factors and Ergonomics Society, 41*(2):312–325. Available: http://hfs.sagepub.com/content/41/2/312.full.pdf+htm [October 2014].

Epstein, S. (2011). Measuring success: Scientific, institutional, and cultural effects of patient advocacy. In B. Hoffman, N. Tomes, R. Grobe, and M. Schlesinger (Eds.), *Patients as Policy Actors: A Century of Changing Markets and Missions*. Piscataway, NJ: Rutgers University Press.

Erdogan, B., and Bauer, T.N. (2010). Differentiated leader-member exchanges: The buffering role of justice climate. *Journal of Applied Psychology, 95*(6):1104–1120.

Espinosa J.A., Cummings, J.N., Wilson, J.M., and Pearce, B.M. (2003). Team boundary issues across multiple global firms. *Journal of Management Information Systems, 19*:157–190.
Falk-Krzesinski, H.J., Contractor, N., Fiore, S.M., Hall, K.L., Kane, C., Keyton, J., et al. (2011). Mapping a research agenda for the science of team science. *Research Evaluation, 20*(2):143–156. Available: http://www.socialresearchmethods.net/research/Mapping%20 a%20research%20agenda%20for%20the%20Science%20of%20Team%20Science.pdf [May 2014].
Feist, G. (2011). Creativity in science. In M.A. Runco and S.R. Pritzker (Eds.), *Encyclopedia of Creativity* (2nd ed., vol. 1, pp. 296–302). London: Elsevier.
Feist, G.J. (2013). Creative personality. In E.G. Carayannis (Ed.), *Encyclopedia of Creativity, Invention, Innovation and Entrepreneurship* (pp. 344–349). New York: Springer.
Feller, I., Gamota, G., and Valdez, W. (2003). Developing science indicators for basic science offices within mission agencies. *Research Evaluation, 12*(1):71–79.
Festinger, L. (1950). Informal social communication. *Psychological Review, 57*:271–282
Finholt, T.A. and Olson, G. (1997). From laboratories to collaboratories: A new organizational form for scientific collaboration. *Psychological Science, 8*(1):28–36.
Fiore, S.M. (2008). Interdisciplinarity as teamwork: How the science of teams can inform team science. *Small Group Research, 39*(3):251–277. Available: http://dx.doi.org/10.1177/1046496408317797 [May 2014].
Fiore, S.M. (2013). *Overview of the Science of Team Science.* Presented at the National Research Council's Planning Meeting on Interdisciplinary Science Teams, January 11, Washington, DC. Available: http://tvworldwide.com/events/nas/130111/ppt/Fiore%20 FINAL%20SciTS%20Overview%20for%20NRC.pdf [May 2014].
Fiore, S.M., and Bedwell, W. (2011). *Team Science Needs Teamwork Training.* Presented at the the Second Annual Science of Team Science Conference, Chicago, IL.
Fiore, S.M., Rosen, M.A., Smith-Jentsch, K.A., Salas, E., Letsky, M. and Warner, N. (2010a). Toward an understanding of macrocognition in teams: Predicting processes in complex collaborative contexts. *Human Factors, 52*(2):203–224.
Fiore, S.M., Smith-Jentsch, K.A., Salas, E., Warner, N., and Letsky, M. (2010b). Toward an understanding of macrocognition in teams: Developing and defining complex collaborative processes and products. *Theoretical Issues in Ergonomic Science, 11*(4):250–271.
Flaherty, C. (2014). Mentor or risk rejection. *Inside Higher Education.* Available: http://www.insidehighered.com/news/2014/06/24/scientists-note-nsf-push-data-mentoring-grant-proposals#sthash.53MTa6jD.rGDMQsI4.dpbs [June 2014].
Forsyth, D.R. (2010). *Group Dynamics.* Belmont, CA: Wadsworth.
Foster, I., and Kesselman, C. (Eds.). (2004). *The Grid: Blueprint for a New Computing Infrastructuxre* (2nd. ed.). San Francisco: Morgan Kaufmann.
Fouse, S., Cooke, N.J., Gorman, J.C., Murray, I., Uribe, M., and Bradbury, A. (2011). Effects of role and location switching on team performance in a collaborative planning environment. *Proceedings of the 55th Annual Conference of the Human Factors and Ergonomics Society, 55*:1442–1446.
Fowlkes, J.E., Lane, N.E., Salas, E., Franz, T., and Oser, R. (1994). Improving the measurement of team performance: The TARGETS Methodology. *Military Psychology, 6*(1):47–61.
Freeman, R.B., and Huang, W. (2014). *Collaborating with People Like Me: Ethnic Co-Authorship within the U.S.* NBER Working Paper No. 19905. Cambridge, MA: National Bureau of Economic Research. Available: http://www.nber.org/papers/w19905 [May 2015].
Frickel, S., and Jacobs, J.J. (2009). Interdisciplinarity: A critical assessment. *American Review of Sociology, 35*:43–65. Available: https://proseminarcrossnationalstudies.files.wordpress.com/2009/11/interdisciplinarity_ars_2009.pdf [April 2015].

Friesenhahn, I., and Beaudry, C. (2014). *The Global State of Young Scientists–Project Report and Recommendations*. Berlin: Akademie Verlag. Available: http://www.raje.es/en/wp-content/uploads/2014/01/GYA_GloSYS-report_webversion.pdf [May 2014].

Frische, S. (2012). It is time for full disclosure of author contributions. *Nature, 489*:475. Available: http://www.nature.com/news/it-is-time-for-full-disclosure-of-author-contributions-1.11475 [May 2014].

Frodeman, R., Klein, J.T., Mitcham, C., and Holbrook, J.B. (2010). *The Oxford Handbook of Interdisciplinarity*. Oxford, UK: Oxford University Press.

Furman, J.L., and Gaule, P. (2013). *A Review of Economic Perspectives on Collaboration in Science*. Prepared for the National Research Council's Workshop on Institutional and Organizational Supports for Team Science, October 24, Washington, DC. Available: http://sites.nationalacademies.org/DBASSE/BBCSS/DBASSE_085357 [May 2014].

Fussell, S.R., and Setlock, L.D. (2012). Multicultural teams. In W.S. Bainbridge (Ed.), *Leadership in Science and Technology: A Reference Handbook* (vol. 1, pp. 255–263). Thousand Oaks, CA: Sage.

Gabelica, C., and Fiore, S.M. (2013a). What can training researchers gain from examination of methods for active-learning (PBL, TBL, and SBL). *Proceedings of the Human Factors and Ergonomics Society Annual Meeting, 57*(1):1462–1466. Available: http://pro.sagepub.com/content/57/1/462.refs [May 2014].

Gabelica, C., and Fiore, S.M. (2013b). *Learning How to Be a (Team) Scientist*. Presented at the 4th Annual Science of Team Science Conference. June 24–27, Northwestern University, Evanston, IL.

Galison, P. (1996). Computer simulations and the trading zone. In P. Galison and D.J. Stump (Eds.), *The Disunity of Science: Boundaries, Contexts, and Power* (pp. 118–157). Stanford, CA: Stanford University Press.

Gallupe, R.B., Dennis, A.R., Cooper, W.H., Valacich, J.S., Bastianutti, L.M. and Nunamaker, J.F. (1992). Electronic brainstorming and group size. *Academy of Management Journal, 35*(2):350-369.

Gans, J.S., and Murray, F. (2015). The changing nature of scientific credit. In A. Jaffe and B. Jones (Eds.), *The Changing Frontier: Rethinking Science and Innovation Policy*. Chicago: University of Chicago Press.

Garrett-Jones, S., Turpin, T., and Diment, K. (2010). Managing competition between individual and organizational goals in cross-sector research and development centres. *The Journal of Technology Transfer, 35*(5):527–546.

Gebbie, K.M., Meier, B.M., Bakken, S., Carrasquillo, O., Formicola, A., Aboelela, S.W., Glied, S., and Larson, E. (2007). Training for interdisciplinary health research: Defining the required competencies. *Journal of Allied Health, 37*(2):65–70. Available: http://ww.michr.umich.edu/Uploads/Education/Training%20for%20Interdisciplinary%20Health%20Research_Defining%20the%20Required%20Competencies.pdf [October 2014].

Gehlert, S., Hall, K.L., Vogel, A.L., Hohl, S., Hartman, S., Nebeling, L., et al. (2014). Advancing transdisciplinary research: The Transdisciplinary Energetics and Cancer Initiative. *The Journal of Translational Medicine and Epidemiology, 2*(2):1032. Available: http://www.jscimedcentral.com/TranslationalMedicine/translationalmedicine-spid-collaboration-science-translational-medicine-1032.pdf [April 2015].

Gerstner, C.R., and Day, D.V. (1997). Meta-analytic review of leader-member exchange theory: Correlates and construct issues. *Journal of Applied Psychology, 82*(6):827–844.

Gibson, C., and Vermeulen, F. (2003). A healthy divide: Subgroups as a stimulus for team learning behavior. *Administrative Science Quarterly, 48*:202–239.

Gibson, C.B., and Gibbs, J.L. (2006). Unpacking the concept of virtuality: The effects of geographic dispersion, electronic dependence, dynamic structure, and national diversity on team innovation. *Administrative Science Quarterly, 51*:451–495. Available: http://www.crito.uci.edu/papers/2006/VirtualInnovation.pdf [October 2014].

Gijbels, D., Dochy, F., Van den Bossche, P., and Segers, M. (2005). Effects of problem-based learning: A meta-analysis from the angle of assessment. *Review of Educational Research, 75*(1):27–61. Available: http://rer.sagepub.com/content/75/1/27.full.pdf [October 2014].

Gladstein, D.L. (1984). Groups in context: A model of task group effectiveness. *Administrative Science Quarterly, 29*(4):499–517.

Gorman, J.C., and Cooke, N.J. (2011). Changes in team cognition after a retention interval: The benefits of mixing it up. *Journal of Experimental Psychology: Applied, 17*(4):303–319.

Gorman, J.C., Amazeen, P.G., and Cooke, N.J. (2010). Team coordination dynamics. *Nonlinear Dynamics Psychology and Life Sciences, 14*:265–289.

Gorman, J.C., Cooke, N.J., and Amazeen, P.G. (2010). Training adaptive teams. *Human Factors, 52*(2):295–307. Available: http://hfs.sagepub.com/content/52/2/295.full.pdf [October 2014].

Gray, B. (2008). Enhancing transdisciplinary research through collaborative leadership. *American Journal of Preventative Medicine, 35*(2S):S124–S132. Available: http://cancercontrol.cancer.gov/brp/scienceteam/ajpm/enhancingtransdisciplinaryresearchthroughcollaborative leadership.pdf [October 2014].

Gray, D.O. (2008). Making team science better: Applying improvement-oriented evaluation principles to evaluation of cooperative research centers. In C.L.S. Coryn and M. Scriven (Eds.), *New Directions for Evaluation* (Ch. 6, pp. 73–87). Wiley Online Library. Available: http://onlinelibrary.wiley.com/doi/10.1002/ev.262/pdf [May 2015].

Gray, D.O., and Walters, S.G. (Eds.). (1998). *Managing the Industry University Cooperative Research Center: A Guide for Directors and Other Stakeholders.* Columbus, OH: Battelle Press.

Grayson, D.M., and Monk, A.F. (2003). Are you looking at me? Eye contact and desktop video conferencing. *Transactions on Computer-Human Interaction, 10*(3):221–243.

Griffin, J. (1943). *The Fort Ancient Aspect: Its Cultural and Chronological Position in Mississippi Valley Archaeology.* Ann Arbor: University of Michigan Press.

Grinter, R.E. (2000). Workflow systems: Occasions for success and failure. *Computer Supported Cooperative Work, 9*:189–214. Available: http://download.springer.com/static/pdf/493/art%253A10.1023%252FA%253A1008719814496.pdf?auth66=1414098069_e182ebba4e4cef2819d047aafa9e54b2&ext=.pdf [October 2014].

Grudin, J. (1994). Groupware and social dynamics: Eight challenges for developers. *Communications of the ACM, 37*(1):93–105. Available: https://www.lri.fr/~mbl/ENS/CSCW/2013/papers/Grudin-CACM-94.pdf [October 2014].

Grudin, J., and Palen, L. (1995). *Why Groupware Succeeds: Discretion or Mandate?* Presented at the ECSCW 1995, Stockholm, Sweden. Available: http://research.microsoft.com/en-us/um/people/jgrudin/past/papers/ecscw95/ecscw.html [October 2014].

Gruenfeld, D.H., Martorana, P.V., and Fan, E.T. (2000). What do groups learn from their worldliest members? *Organizational Behavior and Human Decisions Processes, 82*(1): 45–59. Available: http://paulmartorana.com/wp-content/uploads/2011/09/Gruenfeld_Martorana_Fan_2000.pdf [October 2014].

Guimera, R., Uzzi, B., Spiro, J., and Amaral, L.A.N. (2005). Team assembly mechanisms determine collaboration network structure and team performance. *Science 308*:697–702. Available: http://www.ncbi.nlm.nih.gov/pmc/articles/PMC2128751/pdf/nihms-34163.pdf [October 2014].

Guinan, E.C., Boudreau, K.C., and Lakhani, K.R. (2013). Experiments in open innovation at Harvard Medical School. *M.I.T. Sloan Management Review, 54*(3):45–52.

Gulati, R. (1995). Social structure and alliance formation patterns: A longitudinal analysis. *Administrative Science Quarterly, 40*(4):619–652.

Gully, S.M., Devine, D.J., and Whitney, D.J. (1995). A meta-analysis of cohesion and performance: Effects of levels of analysis and task interdependence. *Small Group Research*, 26(4):497–520. Available: http://sgr.sagepub.com/content/26/4/497.full.pdf [October 2014].

Gully, S.M., Incalcaterra, K.A., Joshi, A., and Beaubien, J.M. (2002). A meta-analysis of team-efficacy, potency, and performance: Interdependence and level of analysis as moderators of observed relationships. *Journal of Applied Psychology*, 87(5):819–832.

Gurtner, A., Tschan, F., Semmer, N.K., and Nägele, C. (2007). Getting groups to develop good strategies: Effects of reflexivity interventions on team process, team performance, and shared mental models. *Organizational Behavior and Human Decision Processes*, 102(2):127–142.

Hackett, E.J. (2005). Essential tensions: Identity, control, and risk in research. *Social Studies of Science*, 35(5):787–826. Available: http://sss.sagepub.com/content/35/5/787.full.pdf [October 2014].

Hackman, J.R. (2012). From causes to conditions. Group influences on individuals in organizations. *Journal of Organizational Behavior*, 33:428–444. Available: http://onlinelibrary.wiley.com/doi/10.1002/job.1774/pdf [October 2014].

Hackman, J.R., and Vidmar, N. (1970). Effects of size and task type on group performance and member reactions. *Sociometry*, 33:37–54.

Hadorn, G., and Pohl, C. (2007). *Principles for Designing Transdisciplinary Research: Proposed by the Swiss Academy of Arts and Sciences*. Available: http://www.transdisciplinarity.ch/documents/knowledgeforms_principles.pdf [October 2014].

Hagstrom, W.O. (1964). *The Scientific Community*. London and New York: Basic Books.

Haines, J.K., Olson, J.S., and Olson, G.M. (2013). *Here or There? How Configurations of Transnational Teams Impact Social Capital*. Presented at Interact 2013. Available: http://www.irit.fr/recherches/ICS/events/conferences/interact2013/papers/8118481.pdf [October 2014].

Hall, D.J., and Saias, M.A. (1980). Strategy follows structure! *Strategic Management Journal*, 1(2):149–163.

Hall, K.L., Stokols, D., Moser, R.P., Taylor, B.K., Thornquist, M.D., Nebeling, L.C., et al. (2008). The collaboration readiness of transdisciplinary research teams and centers: Findings from the National Cancer Institute's TREC Year-One Evaluation Study. *American Journal of Preventive Medicine, Supplement on the Science of Team Science*, 35(2S):S161–S172.

Hall, K.L., Vogel, A.L., Stipelman, B.A., Stokols, D., Morgan, G., and Gehlert, S. (2012a). A four-phase model of transdisciplinary team-based research: Goals, team processes, and strategies. *Translational Behavioral Medicine*, 2(4):415–430.

Hall, K.L., Stokols, D., Stipelman, B., Vogel, A., Feng, A., Masimore, B., et al. (2012b). Assessing the value of team science: A study comparing center- and investigator-initiated grants. *American Journal of Preventive Medicine*, 42(2):157–163.

Hall, K.L., Olster, D., Stipelman, B., and Vogel, A.L. (2012c). News from NIH: Resources for team-based research to more effectively address complex public health problems. *Translational Behavioral Medicine*, 2(4):373–375. Available: http://www.ncbi.nlm.nih.gov/pmc/articles/PMC3717922/pdf/13142_2012_Article_172.pdf [October 2014].

Hall, K.L., Vogel, A.L., Ku, M.C., Klein, J.T., Banacki, A., Bennett, L.M., et al. (2013). *Recognition for Team Science and Cross-disciplinarity in Academia: An Exploration of Promotion and Tenure Policy and Guideline Language from Clinical and Translational Science Awards (CTSA) Institution*. Presented at the National Research Council Workshop on Institutional and Organizational Supports for Team Science, October 24, Washington, DC. Available: http://sites.nationalacademies.org/DBASSE/BBCSS/DBASSE_085357 [March 2014].

Hall, K.L., Crowston, K., and Vogel, A.L. (2014). *How to Write a Collaboration Plan*. Rockville, MD: National Cancer Institute. Available: https://www.teamsciencetoolkit.cancer.gov/Public/TSResourceBiblio.aspx?tid=3&rid=3119 [April 2015].

Hall, K.L., Stipelman, B., Vogel, A.L., Huang, G., and Dathe, M. (2014). *Enhancing the Effectiveness of Team-based Research: A Dynamic Multi-level Systems Map of Integral Factors in Team Science*. Presented at the Fifth Annual Science of Team Science Conference, August, Austin, TX.

Hammond, R.A. (2009). Complex systems modeling for obesity research. *Prevention of Chronic Disease*, 6(3):A97. Available: http://www.cdc.gov/pcd/issues/2009/jul/pdf/09_0017.pdf [October 2014].

Hand, E. (2010). Citizen science: People power. *Nature*, 466(7307):685–687.

Hannah, S.T., and Parry, K.W. (in press). Leadership in extreme contexts. In D.V. Day (Ed.), *The Oxford Handbook of Leadership and Organizations*. New York: Oxford University Press.

Hannah, S.T., Uhl-Bien, M., Avolio, B.J., and Cavaretta, F.L. (2009). A framework for examining leadership in extreme contexts. *Leadership Quarterly*, 2:897–919. Available: http://digitalcommons.unl.edu/cgi/viewcontent.cgi?article=1038&context=managementfacpub [October 2014].

Heffernan, J.B., Sorrano, P.A., Angilletta, M.J., Buckley, L., Gruner, D., Keitt, T. et al. (2014). Macrosystems ecology: Understanding ecological patterns and processes at continental scales. *Frontiers in Ecology*, 12:5–14. Available: http://www.esajournals.org/doi/pdf/10.1890/130017 [January 2014].

Helmreich, R.L., Merritt, A.C., and Wilhelm, J.A. (1999). The evolution of Crew Resource Management training in commercial aviation. *International Journal of Aviation Psychology*, 9(1):19–32.

Hempel, P.S., Zhang, Z.X., and Han, Y. (2012). Team empowerment and the organizational context: Decentralization and the contrasting effects of formalization. *Journal of Management*, 38(2):475–501.

Herbsleb, J.D., and Grinter, R.E. (1999). Architectures, coordination, and distance: Conway's law and beyond. *IEEE Software*, Sept–Oct:63–70. Available: http://www.cs.cmu.edu/afs/cs/Web/People/jdh/collaboratory/research_papers/IEEE99_final.pdf [October 2014].

Hess, D.J. (1997). *Science Studies: An Advanced Introduction*. New York: New York University Press.

Heuer, R.J. (1999). *Psychology of Intelligence Analysis*. Commissioned by the Central Intelligence Agency, Center for the Study of Intelligence.

Hinds, P., and McGrath, C. (2006). *Structures That Work: Social Structure, Work Structure, and Performance in Geographically Distributed Teams*. Available: http://web.stanford.edu/~phinds/PDFs/Hinds-McGrath-2006-CSCW.pdf [October 2014].

Hinnant, C.C., Stvilia, B., Wu, S., Worrall, A., Burnett, G., Burnett, K., et al. (2012). Author-team diversity and the impact of scientific publications: Evidence from physics research at a national science lab. *Library & Information Science Research*, 34:249–257. Available: http://diginole.lib.fsu.edu/cgi/viewcontent.cgi?article=1006&context=slis_faculty_publications [October 2014].

Hinsz, V.B., Tindale, R.S., and Vollrath, D.A. (1997). The emerging conceptualization of groups as information processors. *Psychological Bulletin*, 121(1):43–64.

Hoch, J.E., and Duleborhn, J.H. (2013). Shared leadership in enterprise resource planning and human resource management system implementation. *Human Resource Management Review*, 23(1):114–125.

Hoch, J.E., and Kozlowski, S.W.J. (2014). Leading virtual teams: Hierarchical leadership, structural supports, and shared team leadership. *Journal of Applied Psychology*, 99(3):390–403.

Hodgson, G.M. (2006). What are institutions? *Journal of Economic Issues.* Available: http://www.geoffrey-hodgson.info/user/bin/whatareinstitutions.pdf [March 2015].

Hofer, E. C., McKee, S., Brinholtz, J. P., and Avery, P. (2008) High-energy physics: The large hadron collider collaborations. In G. Olson, N. Bos, and A. Zimmerman (Eds.), *Scientific Collaboration on the Internet.* Cambridge, MA: MIT Press.

Hofmann, D.A., and Jones, L.M. (2005). Leadership, collective personality, and performance. *Journal of Applied Psychology, 90*(3):509–522.

Hofmann, D.A., Morgeson, F.P., and Gerras, S.J. (2003). Climate as a moderator of the relationship between leader-member exchange and content specific citizenship: Safety climate as an exemplar. *Journal of Applied Psychology, 88*(1):170–178. Available: https://www.msu.edu/~morgeson/hofmann_morgeson_gerras_2003.pdf [October 2014].

Hogan, R., Curphy, G.J., and Hogan, J. (1994). What we know about leadership: Effectiveness and personality. *American Psychologist, 49*(6):493–504. Available: http://peterberry.sitesuite.ws/files/hogan_research_articles/journal_articles/what_we_know_about_leadership.pdf [October 2014].

Hogg, M.A., van Knippenberg, D., and Rast, D.E. III. (2012). Intergroup leadership in organizations: Leading across group and organizational boundaries. *Academy of Management Review, 37*(2):232–255.

Hohle, B.M., McInnis, J.K., and Gates, A.C. (1969). The public health nurse as a member of the interdisciplinary team. *The Nursing Clinics of North America, 4*(2):311–319.

Holbrook, J.B. (2010). Peer review. In R. Frodeman, J.T. Klein, and C. Mitcham (Eds.), *The Oxford Handbook of Interdisciplinarity* (pp. 321–332). Oxford, UK: Oxford University Press.

Holbrook, J.B. (2013). *Peer Review of Team Science Research.* Presented at the Workshop on Institutional and Organizational Supports for Team Science, National Research Council, Washington, DC. Available: http://sites.nationalacademies.org/DBASSE/BBCSS/DBASSE_085357 [April 2014].

Holbrook, J.B., and Frodeman, R. (2011). Peer review and the ex-ante assessment of societal impacts. *Research Evaluation, 20*(3):239–246.

Holland, J.H. (1992). Complex adaptive systems. *Daedalus, 12*(1):17–30. Available: http://www.personal.umich.edu/~samoore/bit885w2012/ComplexAdaptiveSystemsHolland.pdf [May 2014].

Hollander, E.P. (1964). *Leaders, Groups, and Influence.* New York: Oxford University Press.

Hollenbeck, J.R., DeRue, D.S., and Guzzo, R. (2004). Bridging the gap between I/O research and HR practice: Improving team composition, team training and team task design. *Human Resource Management, 43*(4):353–366. Available: http://onlinelibrary.wiley.com/doi/10.1002/hrm.20029/pdf [October 2014].

Hollingshead, A.B. (1998). Communication, learning, and retrieval in transactive memory systems. *Journal of Experimental Social Psychology, 34*:423–442.

Holt, V.C. (2013). *Graduate Education to Facilitate Interdisciplinary Research Collaboration: Identifying Individual Competencies and Developmental Learning Activities.* Presented at the Science of Team Science Conference Session on Learning and Training for Team Science, June, Evanston, IL. Available: http://www.scienceofteamscience.org/2013-sessions--learning-and-training-for-team-science [May 2014].

Homan, A.C., Hollenbeck, J.R., Humphrey, S.E., van Knippenberg, D., Ilgen, D.R., and Van Kleef, G.A. (2008). Facing differences with an open mind: Openness to experience, salience of intra-group differences, and performance of diverse work groups. *Academy of Management Journal, 51*(6):1204–1222. Available: http://www.personal.psu.edu/seh25/HomanEtAl2008.pdf [October 2014].

Horstman, T., and Chen, M. (2012). *Gamers as Scientists? The Relationship Between Participating in Foldit Play and Doing Science.* Presented at the American Educational Research Association Annual Meeting. Available: http://www.researchgate.net/publication/258294401_Gamers_as_Scientists_The_Relationship_Between_Participating_in_Foldit_Play_and_Doing_Science [May 2014].

Horwitz, S.K., and Horwitz, I.W. (2007). The effects of team diversity on team outcomes: A meta-analytic review of team demography. *Journal of Management, 33*(6):987–1015.

Howe, J. (2008). *Crowdsourcing: Why the Power of the Crowd Is Driving the Future of Business.* New York: Crown.

Hunt, D.P., Haidet, P., Coverdale, J.H., and Richards, B. (2003). The effect of using team learning in an evidence-based medicine course for medical students. *Teaching and Learning in Medicine: An International Journal, 15*(2):131–139.

Huutoniemi, K., and Tapio, P. (2014). *Transdisciplinary Sustainability Studies: A Heuristic Approach.* Milton Park, Abingdon, UK: Routledge.

Ilgen, D.R., Hollenbeck, J.R., Johnson, M., and Jundt, D. (2005). Teams in organizations: From input-process-output models to IMOI models. *Annual Review of Psychology, 56*:517–543.

Incandela, J. (2013). *Preliminary Response to Innovation in Scientific Multiteam Systems: Confluent and Countervailing Forces.* Presented at the National Research Council's Workshop on Science Team Dynamics and Effectiveness, July 1, Washington, DC. Available: http://sites.nationalacademies.org/DBASSE/BBCSS/DBASSE_083679 [May 2014].

Institute of Medicine. (1999). *To Err Is Human: Building a Safer Health System.* Washington, DC: National Academy Press. Available: http://www.iom.edu/Reports/1999/to-err-is-human-building-a-safer-health-system.aspx [May 2014].

Institute of Medicine. (2004). *NIH Extramural Center Programs—Criteria for Initiation and Evaluation.* Washington, DC: The National Academies Press.

Institute of Medicine. (2010). *Bridging the Evidence Gap in Obesity Prevention: A Framework to Inform Decision Making.* Washington, DC: The National Academies Press. Available: http://books.nap.edu/openbook.php?record_id=12847 [February 2014].

Institute of Medicine. (2013). *The CTSA Program at NIH: Opportunities for Advancing Clinical and Translational Research.* Washington, DC: The National Academies Press.

Isaacs, E., Walendowski, A., Whittaker, S., Schiano, D.J., and Kamm, C. (2002). *The Character, Functions, and Styles of Instant Messaging in the Workplace.* Presented at the 2002 Conference on Computer Supported Collaborative Work, New York. Available: http://nl.ijs.si/janes/wp-content/uploads/2014/09/isaacsothers02.pdf [October 2014].

Jackson, S.E. Brett, J.F., Sessa, V.I., Cooper, D.M., Julin, J.A., and Peyronnin, K. (1991). Some differences make a difference: Interpersonal dissimilarity and group heterogeneity as correlates of recruitment, promotion, and turnover. *Journal of Applied Psychology, 76*:675–689.

Jackson, S.E., May, K.E., and Whitney, K. (1995). Understanding the dynamics of diversity in decision-making teams. In R.A. Guzzo and E. Salas (Eds.), *Team Decision-Making Effectiveness in Organizations* (pp. 204–261). San Francisco: Jossey-Bass.

Jacobs, J.A. (2014). *In Defense of Disciplines.* Chicago, IL: University of Chicago Press.

Jacobs, J.A., and Frickel, S. (2009). Interdisciplinarity: A critical assessment. *Annual Review of Sociology, 35*:43–65.

Jacobson, S.R. (1974). A study of interprofessional collaboration. *Nursing Outlook, 22*: 751–755.

James, L.R., and Jones, A.P. (1974). Organizational climate: A review of theory and research. *Psychological Bulletin, 81*(12):1096–1112.

James Webb Space Telescope Independent Comprehensive Review Panel. (2010). *Final Report.* Washington, DC: National Aeronautics and Space Administration. Available: http://www. nasa.gov/pdf/499224main_JWST-ICRP_Report-FINAL.pdf [March 2015].

Jankowski, N.W. (Ed.) (2009). *e-Research: Transformation in Scholarly Practice.* New York: Routledge.

Jarvenpaa, S.L., Knoll, K., and Leidner, D.E. (1998). Is anybody out there? Antecedents of trust in global virtual teams. *Journal of Management Information Systems, 14:*29–64.

Jehn, K.A. (1995). A multimethod examination of the benefits and detriments of intragroup conflict. *Administrative Science Quarterly,* 40(2):256–282. Available: http://division. aomonline.org/cm/Award-Winning-Papers/2000-MIA-Jehn-ASQ-1995.pdf [October 2014].

Jehn, K.A. (1997). A qualitative analysis of conflict types and dimensions in organizational groups. *Administrative Science Quarterly,* 42:530–557.

Jehn, K.A., and Bezrukova, K. (2010). The faultline activation process and the effects of activated faultlines on coalition formation, conflict, and group outcomes. *Organizational Behavior and Human Decision Process,* 112(1):24–42.

Jin, G., Jones, B., Feng Lu, S., and Uzzi, B. (2014). *The Reverse Matthew Effect in Teams and Networks.* NBER Working Paper No. 19489. Cambridge, MA: National Bureau of Economic Research. Available: http://www.nber.org/papers/w19489.pdf [April 2015].

Johnson, W.L., and Valente, A. (2009). Tactical language and culture training systems: Using AI to teach foreign languages and cultures. *AI Magazine,* 30(2):72–83.

Jones, B., Wuchty, S., and Uzzi, B. (2008). Multi-university research teams: Shifting impact, geography, and stratification in science. *Science* 322(5905):1259–1262. Available: http:// www.sciencemag.org/content/322/5905/1259.full.pdf [October 2014].

Jones, B.F. (2009). The burden of knowledge and the death of the Renaissance man: Is innovation getting harder? *Review of Economic Studies,* January. Available: http://www. kellogg.northwestern.edu/faculty/jones-ben/htm/burdenofknowledge.pdf [October 2014].

Jordan, G. (2010). A theory-based logic model for innovation policy and evaluation. *Research Evaluation,* 19(4):263–274.

Jordan, G.B. (2013). *A Logical Framework for Evaluating the Outcomes of Team Science.* Presented at the Workshop on Institutional and Organizational Supports for Team Science, National Research Council, Washington, DC. Available: http://sites.nationalacademies. org/DBASSE/BBCSS/DBASSE_085236 [May 2014].

Joshi, A., and Roh, H. (2009). The role of context in work team diversity research: A meta-analytic review. *Academy of Management Journal,* 52(3):599–628.

Judge, T.A., Bono, J.E., Hies, R., and Gerhardt, M.W. (2002). Personality and leadership: A qualitative and quantitative review. *Journal of Applied Psychology,* 87(4):765–780. Available: http://www.timothy-judge.com/Judge,%20Bono,%20Ilies,%20&%20Gerhardt.pdf [October 2014].

Judge, T.A., Piccolo, R.F., and Ilies, R. (2004). The forgotten ones? The validity of consideration and initiating structure in leadership research. *Journal of Applied Psychology,* 89(1):36–51.

Kabo, F., Cotton-Nessler, N., Hwang, Y., Levenstein, M., and Owen-Smith, J. (2013a). *Proximity Effects on the Dynamics and Outcomes of Scientific Collaborations.* University of Michigan Working paper.

Kabo, F., Hwang, Y., Levenstein, M., and Owen-Smith, J. (2013b). Shared paths to the lab: A sociospatial network analysis of collaboration. *Environment and Behavior,* DOI: 10.1177/0013916513493909. Available: http://eab.sagepub.com/content/early/ 2013/07/20/0013916513493909.full.pdf+html [May 2014].

Kahlon, M., Yuan, L., Daigre, J., Meeks, E., Nelson, K., Piontkowski, C., Reuter, K., Sak, R., Turner, B., Webber, G.M., and Chatterjee, A. (2014). The use and significance of a research networking system. *Journal of Medical Internet Research*, 16(2):e46.

Kahn, R.L. (1993). *An Experiment in Scientific Organization*. Chicago: John D. and Catherine T. MacArthur Foundation, Program in Mental Health Development.

Kahn, R.L., and Prager, D.J. (1994). Interdisciplinary collaborations are a scientific and social imperative. *The Scientist* 8(14):12. Available: http://www.the-scientist.com/?articles. view/articleNo/28160/title/Interdisciplinary-Collaborations-Are-A-Scientific-And-Social-Imperative/ [May 2014].

Kantrowitz, T.M. (2005). *Development and Construct Validation of a Measure of Soft Skills Performance*. (Unpublished dissertation). Georgia Institute of Technology, Atlanta, GA.

Karasti, H., Baker, K.S., and Millerant, F. (2010). Infrastructure time: Long-term matters in collaborative development. *Computer Supported Cooperative Work*, 19:377–415. Available: http://download.springer.com/static/pdf/415/art%253A10.1007%252Fs10606-010_9113-z.pdf?auth66=1414414825_872fe5140b195f48247d13b7f2c07370&ext=.pdf [October 2014].

Keller, R.T. (2006). Transformational leadership, initiating structure, and substitutes for leadership: A longitudinal study of research and development project team performance. *Journal of Applied Psychology*, 9(1):202–210. Available: http://umesorld.files.wordpress. com/2011/02/transformational-leadership-initiating-structure-substitutes-for-leadership-keller-r-2006.pdf [October 2014].

Kellogg, K., Orlikowski, W.J., and Yates, J. (2006). Life in the trading zone: Structuring coordination across boundaries in post-bureaucratic organizations. *Organization Science*, 17(1):22–44.

Kelly, R. (1995). *The Foraging Spectrum: Diversity in Hunter-Gatherer Lifeways*. Washington, DC: Smithsonian Institution Press.

Keltner, J.W. (1957). *Group Discussion Processes*. New York: Longmans, Green.

Kendon, A. (1967). Some functions of gaze direction in social interactions. *Acta Psychologica*, 26:22–63.

Kerr, N.L., and Tindale, R.S.. (2004). Group performance and decision making. *Annual Review of Psychology*, 55(1):623–655.

Kezar, A., and Maxey, D. (2013). The changing academic workforce. *Trusteeship*, 21(3). Available: http://agb.org/trusteeship/2013/5/changing-academic-workforce [April 2015].

King, H.B., Battles, J., Baker, D.P., Alonso, A., Salas, E., Webster, J., et al. (2008). Team-STEPPS: Strategies and tools to improve patient safety. In K. Henriksen et al. (Eds.), *Advances in Patient Safety: New Directions and Alternative Approaches* (Vol. 3: Performance and Tools). Rockville, MD: Agency for Health Care Research and Quality. Available: http://www.ahrq.gov/professionals/quality-patient-safety/patient-safety-resources/resources/advances-in-patient-safety-2/index.html [December 2014].

Kirkman, B.L., and Rosen, B. (1999). Beyond self-management: Antecedents and consequences of team empowerment. *Academy of Management Journal*, 42:58–74.

Kirkman, B.L., and Mathieu, J.E. (2005). The dimensions and antecedents of team virtuality. *Journal of Management*, 31(5):700–718. Available: http://jom.sagepub.com/content/31/5/700.full.pdf [October 2014].

Kirkman, B.L., Gibson, C.B., and Kim, K. (2012). Across borders and technologies: Advancements in virtual teams research. In S.W.J. Kozlowski (Ed.), *The Oxford Handbook of Organizational Psychology*. New York: Oxford University Press.

Kirwan, B., and Ainsworth, L. (Eds.) (1992). *A Guide to Task Analysis*. London: Taylor and Francis.

Klein, C., DeRouin, R.E., and Salas, E. (2006). Uncovering workplace interpersonal skills: A review, framework, and research agenda. In G.P. Hodgkinson and J.K. Ford (Eds.), *International Review of Industrial and Organizational Psychology* (vol. 21, pp. 80–126). New York: John Wiley & Sons.

Klein, C., DiazGranados, D., Salas, E., Huy, L., Burke, C.S., Lyons, R., and Goodwin, G.F. (2009). Does team building work? *Small Group Research, 40*(2):181–222. Available: http://sgr.sagepub.com/content/40/2/181.full.pdf [October 2014].

Klein, J.T. (1996). *Crossing Boundaries: Knowledge Disciplinarities, and Interdisciplinarities.* Charlottesville: University of Virginia Press.

Klein, J.T. (2010). *Creating Interdisciplinary Campus Cultures: A Model for Strength and Sustainability.* San Francisco: Jossey-Bass.

Klein, J.T., Banaki, A., Falk-Krzesinski, H., Hall, K., Michelle Bennett, L.M. and Gadlin, H. (2013). *Promotion and Tenure in Interdisciplinary Team Science: An Introductory Literature Review.* Presented at the National Research Council Workshop on Organizational and Institutional Supports for Team Science, Washington, DC. Available: http://sites.nationalacademies.org/DBASSE/BBCSS/DBASSE_085357 [May 2014].

Kleingeld, A., van Mierlo, H., and Arends, L. (2011). The effect of goal setting on group performance: A meta-analysis. *Journal of Applied Psychology, 96*(6):1289–1304.

Klimoski, R.J., and Jones, R.G. (1995). Staffing for effective group decision making: Key issues in matching people and teams. In R.A. Guzzo and E. Salas (Eds.), *Team Effectiveness and Decision Making in Organizations* (pp. 291–332). San Francisco: Jossey-Bass.

Knorr, K.D., Mittermeir, R., Aichholzer, G., and Waller, G. (1979). Leadership and group performance: A positive relationship in academic research units. In F.M. Andrews (Ed.), *Scientific Productivity: The Effectiveness of Research Groups in Six Countries.* Cambridge, UK: Cambridge University Press.

Knorr-Cetina, K. (1999). *Epistemic Cultures: How the Sciences Make Knowledge.* Cambridge, MA: Harvard University Press.

Koehne, G., Shih, P.C., and Olson, J.S. (2012). Remote and alone: Coping with being the remote member on the team. In *Proceedings of the ACM Conference on Computer Supported Cooperative Work* (pp. 1257–1266). New York: ACM.

Kotter, J.P. (2001). What leaders really do. *Harvard Business Review, 79*(11):85–96. Available: http://ag.udel.edu/longwoodgrad/symposium/2014/pdf/What%20Leaders%20Really%20Do.pdf [October 2014].

Kozlowski, S.W.J. (2012). Groups and teams in organizations: Studying the multilevel dynamics of emergence. In A.B. Hollingshead and M.S. Poole (Eds.), *Research Methods for Studying Groups and Teams: A Guide to Approaches, Tools, and Technologies* (pp. 260–283). New York: Routledge.

Kozlowski, S.W.J. (in press). Advancing research on team process dynamics: Theoretical, methodological, and measurement considerations. *Organizational Psychology Review.* Available: http://iopsych.msu.edu/koz/Recent%20Pubs/Kozlowski%20%28in%20press%29%20-%20Team%20Dynamics.pdf [May 2015].

Kozlowski, S.W.J., and Bell, B.S. (2003). Work groups and teams in organizations. In W.C. Borman, D.R. Ilgen, and R.J. Kilmoski (Eds.), *Handbook of Psychology: Industrial and Organizational Psychology* (vol. 12, pp. 333–375). London: John Wiley & Sons.

Kozlowski, S.W.J., and Bell, B.S. (2013). Work groups and teams in organizations: Review update. In N. Schmitt and S. Highhouse (Eds.), *Handbook of Psychology: Industrial and Organizational Psychology* (vol. 12, 2nd ed., pp. 412–469). London: John Wiley & Sons. Available: http://digitalcommons.ilr.cornell.edu/cgi/viewcontent.cgi?article=1396&context=articles [October 2014].

Kozlowski, S.W., and Doherty, M.L. (1989). Integration of climate and leadership: Examination of a neglected issue. *Journal of Applied Psychology, 74*(4):546–553.

Kozlowski, S.W.J., and Hults, B.M. (1987). An exploration of climates for technical updating and performance. *Personnel Psychology, 40*:539–563.

Kozlowski, S.W.J., and Ilgen, D.R. (2006). Enhancing the effectiveness of work groups and teams. *Psychological Science in the Public Interest, 7*(3):77–124. Available: http://dx.doi.org/10.1111/j.1529-1006.2006.00030.x [May 2014].

Kozlowski, S.W.J., and Klein, K.J. (2000). A multilevel approach to theory and research in organizations: Contextual, temporal, and emergent processes. In K.J. Klein and S.W.J. Kozlowski (Eds.), *Multilevel Theory, Research and Methods in Organizations: Foundations, Extensions, and New Directions* (pp. 3–90). San Francisco: Jossey-Bass.

Kozlowski, S.W.J., Gully, S.M., McHugh, P.P., Salas, E., and Cannon-Bowers, J.A. (1996). A dynamic theory of leadership and team effectiveness: Developmental and task contingent leader roles. In G.R. Ferris (Ed.), *Research in Personnel and Human Resource Management* (vol. 14, pp. 253–305). Greenwich, CT: JAI Press.

Kozlowski, S.W.J., Gully, S.M., Nason, E.R., and Smith, E.M. (1999). A dynamic theory of leadership and team effectiveness: Developmental and task contingent leader roles. In D.R. Ilgen and E.D. Pulakos (Eds.), *The Changing Nature of Performance: Implications for Staffing, Motivation, and Development* (pp. 240–292). San Francisco: Jossey-Bass.

Kozlowski, S.W.J., Brown, K.G., Weissbein, D.A., Cannon-Bowers, J.A., and Salas, E. (2000). A multi-level perspective on training effectiveness: Enhancing horizontal and vertical transfer. In K.J. Klein and S.W.J. Kozlowski (Eds.), *Multilevel Theory, Research, and Methods in Organizations* (pp. 157–210). San Francisco: Jossey-Bass.

Kozlowski, S.W.J., Watola, D.J., Jensen, J.M., Kim, B.H., and Botero, I.C. (2009). Developing adaptive teams: A theory of dynamic team leadership. In E. Salas, G.F. Goodwin, and C.S. Burke (Eds.), *Team Effectiveness in Complex Organizations: Cross-Disciplinary Perspectives and Approaches* (pp. 113–155). New York: Routledge.

Kozlowski, S.W.J., Chao, G.T., Grand, J.A., Braun, M.T., and Kuljanin, G. (2013). Advancing multilevel research design: Capturing the dynamics of emergence. *Organizational Research Methods, 16*:581–615. Available: http://orm.sagepub.com/content/early/2013/06/24/1094428113493119.full.pdf [October 2014].

Kozlowski, S.W.J., Chao, G.T., Grand, J.A., Braun, M.T., and Kuljanin, G. (in press). Capturing the multilevel dynamics of emergence: Computational modeling, simulation, and virtual experimentation. *Organizational Psychology Review.* Available: http://iopsych.msu.edu/koz/Recent%20Pubs/Kozlowski%20et%20al.%20%28in%20press%29%20-%20Dynamics%20of%20Emergence-Comp%20Modeling,%20Sim,%20Virtual%20Exp.pdf [May 2015].

Kraiger, K., Ford, J.K., and Salas, E. (1993). Application of cognitive, skill-based, and affective theories of learning outcomes to new methods of training evaluation. *Journal of Applied Psychology, 78*(2):311–328. Available: http://www.owlnet.rice.edu/~ajv2/courses/12a_psyc630001/Kraiger,%20Ford,%20&%20Salas%20(1993)%20JAP.pdf [October, 2014].

Kumpfer, K.L., Turner, C., Hopkins, R., and Librett, J. (1993). Leadership and team effectiveness in community coalitions for the prevention of alcohol and other drug abuse. *Health Education Research, 8*(3):359–374.

Lamont, M., and White, P. (2005). *Workshop on Interdisciplinary Standards for Systematic Qualitative Research.* National Science Foundation, Arlington, VA, May 19–20. Available: http://bit.ly/Th5sQp [May 2014].

Latané, B., Williams, K., and Harkins, S. (1979). Many hands make light the work: The causes and consequences of social loafing. *Journal of Personality and Social Psychology, 37*(6):822–832.

Latour, B., and Woolgar, S. (1986). *Laboratory Life: The Construction of Scientific Facts.* Princeton, NJ: Princeton University Press.

Lattuca, L.R., Knight, D., and Bergom, I. (2013). Developing a measure of interdisciplinary competence. *International Journal of Engineering Education, 29*(3):726–739.

Lattuca, L.R., Knight, D.B., Seifert, T., Reason, R.D., and Liu, Q. (2013). *The Influence of Interdisciplinary Undergraduate Programs on Learning Outcomes.* Presented at the 94th annual meeting of the American Educational Research Association, San Francisco, CA.

Lau, D.C., and Murnighan, J.K. (1998). Demographic diversity and faultlines: The compositional dynamics of organizational groups. *Academy of Management Review, 23*(2):325.

Lavin, M.A., Reubling, I., Banks, R., Block, L., Counte, M., Furman, G., et al. (2001). Interdisciplinary health professional education a historical review. *Advances in Health Sciences Education, 6*(1):25–37.

Lawrence, P.R., and Lorsch, J.W. (1967). Differentiation and integration in complex organizations. *Administrative Sciences Quarterly, 12*:1–30.

Lee, J.D., and Kirlik, A. (2013). *The Oxford Handbook of Cognitive Engineering.* New York: Oxford University Press.

LePine, J.A., Piccolo, R. F., Jackson, C.L., Mathieu, J.E., and Saul, J.R. (2008). A meta-analysis of teamwork processes: Tests of a multi-dimensional model and relationships with team effectiveness criteria. *Personnel Psychology, 61*:273–307.

Letsky, M., Warner, N., Fiore, S.M., and Smith, C. (Eds.). (2008). *Macrocognition in Teams: Theories and Methodologies.* London: Ashgate.

Levine, R.A., and Campbell, D.T. (1972). *Ethnocentrism.* New York: John Wiley & Sons.

Lewis, K. (2003). Measuring transactive memory systems in the field: Scale development and validation. *Journal of Applied Psychology, 88*(4):587–604. Available: http://www.owlnet. rice.edu/~ajv2/courses/09a_psyc630001/Lewis%20(2003)%20JAP.pdf [October 2014].

Lewis, K. (2004). Knowledge and performance in knowledge-worker teams: A longitudinal study of transactive memory systems. *Management Science, 50*(11):1519–1533.

Lewis, K., Lange, D., and Gillis, L. (2005). Transactive memory systems, learning, and learning transfer. *Organization Science, 16*(6):581–598.

Lewis, K., Belliveau, M., Herndon, B., and Keller, J. (2007). Group cognition, membership change, and performance: Investigating the benefits and detriments of collective knowledge. *Organizational Behavior and Human Decision Processes, 103*(2):159–178.

Li, J., Ning, Y., Hedley, W., Saunders, B., Chen, Y., Tindill, N., et al. (2002). The molecule pages database. *Nature, 420*:716–717. Available: http://www.nature.com/nature/journal/ v420/n6916/full/nature01307.html [May 2013].

Liang, D.W., Moreland, R.L., and Argote, L. (1995). Group versus individual training and group performance: The mediating role of transactive memory. *Personality and Social Psychology Bulletin, 21*(4):384–393. Available: http://psp.sagepub.com/content/21/4/384. full.pdf [October 2014].

Liden, R.C., Wayne, S.J., Jaworski, R.A., and Bennett, N. (2004). Social loafing: A field investigation. *Journal of Management, 30*(2):285–304.

Liljenström, H., and Svedin, U. (Eds.). (2005). System features, dynamics, and resilience: Some introductory remarks. In H. Liljenström and U. Svedin (Eds.), *Micro–Meso–Macro: Addressing Complex Systems Couplings* (pp. 1–16). London: World Scientific.

Lim, B.C., and Ployhart, R.E. (2004). Transformational leadership: Relations to the five-factor model and team performance in typical and maximum contexts. *Journal of Applied Psychology, 89*(4):610–621.

Lord, R.G., DeVader, C.L, and Alliger, G.M. (1986). A meta-analysis of the relation between personality traits and leadership perceptions: An application of validity generalization procedures. *Journal of Applied Psychology, 71*(3):402–410.

Loughry, M., Ohland, M., and Moore, D. (2007). Development of a theory-based assessment of team member effectiveness. *Educational and Psychological Measurement, 67*(3):505–524. Available: http://epm.sagepub.com/content/67/3/505.full.pdf [October 2014].

Luo, A., Zheng, K., Bhavani, S., and Warden, M. (2010*). Institutional Infrastructure to Support Translational Research.* Presented at the IEEE Sixth International Conference on e-Science, Brisbane, Queensland, Australia, December 7–10. Available: http://doi.ieeecomputersociety.org/10.1109/eScience.2010.50 [May 2015].

Lupella, R.O. (1972). Postgraduate clinical training in speech pathology-audiology: Experiences in an interdisciplinary medical setting. *ASHA, 14*(11):611–614.

Mackay, W.F. (1989). Diversity in the use of electronic mail: A preliminary inquiry. *ACM Transactions on Office Information Systems,* 6:380–397. Available: http://www-ihm.lri.fr/~mackay/pdffiles/TOIS88.Diversity.pdf [Ocotber 2014].

The Madrillon Group, Inc. (2010). *Evaluation of Research Center and Network Programs at the National Institutes of Health: A Review of Evaluation Practice, 1978–2009.* Available: https://www.teamsciencetoolkit.cancer.gov/Public/TSDownload.aspx?aid=176 [October 2014].

Major, D.A., and Kozlowski, S.W.J. (1991). *Organizational Socialization: The Effects of Newcomer, Co-worker, and Supervisor Proaction.* Presented at the Sixth Annual Conference of the Society for Industrial and Organizational Psychology, St. Louis, MO.

Malone, T., and Crowston, K. (1994). The interdisciplinary study of coordination. *ACM Computing Surveys,* 26(1):87–119. Available: http://www.cs.unicam.it/merelli/Calcolo/malone.pdf [October 2014].

Malone, T.W., Laubacher, R., and Dellarocas, C. (2010). The collective intelligence genome. *Sloan Management Review, 51*(3):21–31. Available: http://sloanreview.mit.edu/article/the-collective-intelligence-genome/ [October 2014].

Mann, R.D. (1959). A review of the relationship between personality and performance in small groups. *Psychological Bulletin, 56*(4):241–270.

Mannix, E., and Neale, M.A. (2005). What differences make a difference? The promise and reality of diverse teams in organizations. *Psychological Science in the Public Interest,* 6:31–55. Available: http://psi.sagepub.com/content/6/2/31.full.pdf [October 2014].

Marks, M.A., Zaccaro, S.J., and Mathieu, J.E. (2000). Performance implications of leader briefings and team-interaction training for team adaptation to novel environments. *Journal of Applied Psychology,* 85(6):971–986.

Marks, M.A., Mathieu, J.E., and Zaccaro, S.J. (2001). A temporally based framework and taxonomy of team processes. *Academy of Management Review,* 26(3):356–376. Available: http://www.owlnet.rice.edu/~ajv2/courses/12c_psyc438001/Marks%20et%20al.%20(2001).pdf [October 2014].

Marks, M.A., Sabella, M.J., Burke, C.S., and Zaccaro, S.J. (2002). The impact of cross-training on team effectiveness. *Journal of Applied Psychology,* 87(1):3–13.

Martinez, F.D. (2013). *Faculty Issues: A Matter of Leadership and Governance.* Presented at the National Research Council Workshop on Key Challenges in the Implementation of Convergence, September 16–17, Washington, DC. Available: http://dels.nas.edu/Past-Events/Workshop-Challenges/DELS-BLS-11-08/6665 [May 2014].

Martins, L.L., Gilson, L.L., and Maynard, M.T. (2004). Virtual teams: What do we know and where do we go from here? *Journal of Management,* 30(6):805–835. Available: http://www.owlnet.rice.edu/~ajv2/courses/12c_psyc438001/Martins%20et%20al.%20(2004).pdf [October 2014].

Massey, A.P., Montoya-Weiss, M.M., and Hung, Y.T. (2003). Because time matters: Temporal coordination in global virtual project teams. *Journal of Management Information Systems,* 19(4):129–155.

Mathieu, J.E., and Rapp, T.L. (2009). Laying the foundation for successful team performance trajectories: The roles of team charters and performance strategies. *Journal of Applied Psychology,* 94(1):90–103. Available: https://organized-change-consultancy.wikispaces.com/file/view/Laying+the+Foundation+for+Successful+Team+Performance+Trajectories.pdf [October 2014].

Mathieu, J.E., Heffner, T.S., Goodwin, G.F., Salas, E., and Cannon-Bowers, J.A. (2000). The influence of shared mental models on team process and performance. *Journal of Applied Psychology, 85*(2):273–283.Available: https://www.ida.liu.se/~729A15/mtrl/shared_mental-models_mathieu.pdf [October 2014].

Mathieu, J., Maynard, M., Rapp, T., and Gilson, L. (2008). Team effectiveness 1997–2007: A review of recent advancements and a glimpse into the future. *Journal of Management, 34*(3):410–476. Available: http://dx.doi.org/ 10.1177/0149206308316061 [May 2014].

Mathieu, J.E., Tannenbaum, S.I., Donsbach, J.S., and Alliger, G.M. (2014). A review and integration of team composition models: Moving toward a dynamic and temporal framework. *Journal of Management, 40*(1):130–160. Available: http://jom.sagepub.com/content/40/1/130.full.pdf [October 2014].

Maynard, T., Mathieu, J.E., Gilson, L., and Rapp, T. (2012). Something(s) old and something(s) new: Modeling drivers of global virtual team effectiveness. *Journal of Organizational Behavior, 33*:342–365.

McCann, C., Baranski, J.V., Thompson, M.M., and Pigeau, R. (2000). On the utility of experiential cross-training for team decision making under time stress. *Ergonomics, 43*(8):1095–1110.

McCrae, R.R., and Costa, P.T. (1999). A five-factor theory of personality. In O.P. John, R.W. Robins, and L.A. Pervin (Eds.), *Handbook of Personality: Theory and Research.* New York: Guilford Press.

McDaniel, S.E., Olson, G.M., and Magee, J.C. (1996). Identifying and analyzing multiple threads in computer-mediated and face-to-face conversations. In *Proceedings of the Conference on Computer Supported Cooperative Work* (pp. 39–47). New York: ACM.

McGrath, J.E. (1964). *Social Psychology: A Brief Introduction.* New York: Holt, Rinehart, and Winston.

Merton, R.K. (1968). The Matthew Effect in science. *Science, 159*(3810):56–63. Available: http://www.garfield.library.upenn.edu/merton/matthew1.pdf [October 2014].

Merton, R.K. (1988). The Matthew Effect in science, II: Cumulative advantage and the symbolism of intellectual property. *Isis, 79*(4):606–623. Available: http://garfield.library.upenn.edu/merton/matthewii.pdf [May 2015].

Mintzberg, H. (1990). The design school: Reconsidering the basic premises of strategic management. *Strategic Management Journal, 11*(13):171–195.

Miron-spektor, E., Erez, M., and Naveh, E. (2011). The effect of conformist and attentive-to-detail members on team innovation: Reconciling the innovation paradox. *Academy of Management Journal, 54*(4):740–760.

Misra, R. (2011). R&D team creativity: A way to team innovation. *Ubit, 4*(2):31–35.

Misra, S., Harvey, R.H., Stokols, D., Pine, K.H., Fuqua, J., Shokair, S., and Whiteley, J. (2009). Evaluating an interdisciplinary undergraduate training program in health promotion research. *American Journal of Preventive Medicine, 36*(4):358–365.

Misra, S., Stokols, D., Hall, K.L., and Feng, A. (2011a). Transdisciplinary training in health research: Distinctive features and future directions. In M. Kirst, N. Schaefer-McDaniel, S. Hwang, and P. O'Campo (Eds.), *Converging Disciplines: A Transdisciplinary Research Approach to Urban Health Problems* (pp. 133–147). New York: Springer.

Misra, S., Stokols, D., Hall, K.L., Feng, A., and Stipelman, B.A. (2011b). Collaborative processes in transdisciplinary research and efforts to translate scientific knowledge into evidence-based health practices and policies. In M. Kirst, N. Schaefer-McDaniel, S. Hwang, and P. O'Campo (Eds.), *Converging Disciplines: A Transdisciplinary Research Approach to Urban Health Problems* (pp. 97–110). New York: Springer.

Mitrany, M., and Stokols, D. (2005). Gauging the transdisciplinary qualities and outcomes of doctoral training programs. *Journal of Planning Education and Research,* 24:437–449. Available: https://webfiles.uci.edu/dstokols/Pubs/Mitrany%20%26%20Stokols%20 JPER.pdf [October 2014].

Mohammed, S., Klimoski, R., and Rentsch, J.R. (2000). The measurement of team mental models: We have no shared schema. *Organizational Research Methods,* 3:123–165. Available: http://orm.sagepub.com/content/3/2/123.full.pdf [October 2014].

Mohammed, S., Ferzandi, L., and Hamilton, K. (2010). Metaphor no more: A 15-year review of the team mental model construct. *Journal of Management,* 36(4):876–910. Available: http://jom.sagepub.com/content/early/2010/02/11/0149206309356804.full.pdf [October 2014].

Morgeson, F.P., DeRue, D.S., and Peterson, E. (2010). Leadership in teams: A functional approach to understanding leadership structures and processes. *Journal of Management,* 36(1):5–39. Available: https://www.msu.edu/~morgeson/morgeson_derue_karam_2010. pdf [October 2014].

Mote, J., Jordan, G., and Hage, J. (2007). Measuring radical innovation in real time. *International Journal of Technology, Policy, and Management,* 7(4):355–377.

Muller, M.J., Raven, M.E., Kogan, S., Millen, D.R., and Carey, K. (2003). Introducing chat into business organizations: Toward an instant messaging maturity model. In *GROUP '03: Proceedings of the 2003 International ACM SIGGROUP Conference on Supporting Group Work* (pp. 50–57). New York: ACM.

Mullins, N.C. (1972). The development of a scientific specialty: The phage group and the origins of molecular biology. *Minerva,* 10(1):51–82.

Murayama, K., Matsumoto, M., Izuma, K., and Matsumoto, K. (2010). Neural basis of the undermining effect of monetary reward on intrinsic motivation. *Proceedings of the National Academy of Sciences of the United States of America,* 107(49): 20911–20916. Available: http://www.pnas.org/content/107/49/20911.full.pdf+html [April 2015].

Murphy, E. (2013). *Response to Bienen and Jacobs.* Presented at the National Research Council Workshop on Institutional and Organizational Supports for Team Science, October, Washington, DC. Available: http://sites.nationalacademies.org/DBASSE/BBCSS/ DBASSE_085357 [October 2014].

Murphy, S.N., Dubey, A., Embi, P.J., Harris, P.A., Richter, B.G., Turisco, F., et al. (2012). Current state of information technologies for the clinical research enterprise across academic medical centers. *Clinical and Translational Science,* 5(3):281–284. Available: http:// onlinelibrary.wiley.com/doi/10.1111/j.1752-8062.2011.00387.x/pdf [October 2014].

Murray, F.E. (2012). *Evaluating the Role of Science Philanthropy in American Research Universities.* NBER Working Paper No. 18146. Cambridge, MA: National Bureau of Economic Research. Available: http://www.nber.org/papers/w18146 [May 2014].

Myers, J.D. (2008). A national user facility that fits on your desk: The evolution of collaboratories at the Pacific Northwest National Laboratory. In G.M. Olson, A. Zimmerman, and N. Bos (Eds.), *Scientific Collaboration on the Internet* (pp. 121–134). Cambridge, MA: MIT Press

Nagel, J.D., Koch, A., Guimond, J.M., Galvin, S., and Geller, S. (2013). Building the women's health research workforce: Fostering interdisciplinary research approaches in women's health. *Global Advances in Health and Medicine,* 2(5):24–29. Available: http://orwh. od.nih.gov/news/pdf/gahmj-2013nagel.pdf [October 2014].

Naikar, N., Pearce, B., Drumm, D., and Sanderson, P.M. (2003). Designing teams for first-of-a-kind, complex systems using the initial phases of cognitive work analysis: Case study. *Human Factors,* 45(2):202–217.

Nardi, B.A., Whittaker, S., and Bradner, E. (2000). Interaction and outeraction: Instant messaging in action. In *Proceedings of the 2000 ACM Conference on Computer Supported Cooperative Work* (pp. 79–88). New York: ACM.

Nash, J. (2008). Transdisciplinary training: Key components and prerequisites for success. *American Journal of Preventive Medicine, 35*(2):S133–S140.

Nash, J.M., Collins, B.N., Loughlin, S.E., Solbrig, M., Harvey, R., Krishnan-Sarin, S., et al. (2003). Training the transdisciplinary scientist: A general framework applied to tobacco use behavior. *Nicotine Tobacco Research*, Suppl. 1:S41–S53.

National Academy of Sciences, National Academy of Engineering, and Institute of Medicine. (2005). *Facilitating Interdisciplinary Research*. Committee on Facilitating Interdisciplinary Research and Committee on Science, Engineering, and Public Policy. Washington, DC: The National Academies Press.

National Cancer Institute. (2011). *NCI Team Science Toolkit*. Rockville, MD: Author. Available: https://www.teamsciencetoolkit.cancer.gov/Public/Home.aspx [May 2014].

National Cancer Institute. (2012). *NCI Team Science Workshop, February 7–8, 2012: Summary Notes*. Unpublished manuscript provided by the National Institutes of Health, Rockville, MD.

National Cancer Institute. (2015). *Key Initiatives: NCI Network on Biobehavioral Pathways in Cancer*. Available: http://cancercontrol.cancer.gov/brp/bbpsb/ncintwk-biopthwys.html [April 2015].

National Institutes of Health. (2007). *Enhancing Peer Review at NIH*. Available: http://enhancing-peer-review.nih.gov/meetings/102207-summary.html [May 2014].

National Institutes of Health. (2010). *Collaboration and Team Science*. Available: https://ccrod.cancer.gov/confluence/display/NIHOMBUD/Home [May 2014].

National Institutes of Health. (2011). *Revised Policy: Managing Conflict of Interest in the Initial Peer Review of NIH Grant and Cooperative Agreement Applications*. Available: http://grants.nih.gov/grants/guide/notice-files/NOT-OD-11-120.html [April 2015].

National Institutes of Health. (2013). *Scientific Management Review Board Draft Report on Approaches to Assess the Value of Biomedical Research Supported by NIH*. Available: http://smrb.od.nih.gov/documents/reports/VOBR-Report-122013.pdf [May 2014].

National Research Council. (2006). *America's Lab Report: Investigations in High School Science*. Committee on High School Laboratories: Role and Vision. S.R. Singer, M.L. Hilton, and H.A. Schweingruber (Eds.). Board on Science Education. Center for Education, Division of Behavioral and Social Sciences and Education. Washington, DC: The National Academies Press.

National Research Council. (2007a). *Human-System Integration in the System Development Process: A New Look*. Committee on Human-System Design Support for Changing Technology. R.W. Pew and A.S. Mavor (Eds.). Committee on Human Factors, Division of Behavioral and Social Sciences and Education. Washington, DC: The National Academies Press.

National Research Council. (2007b). *Taking Science to School: Learning and Teaching Science in Grades K–8*. R.A. Duschl, H.A. Schweingruber, and A.W. Shouse (Eds.). Committee on Science Learning, Kindergarten Through Eighth Grade. Board on Science Education. Center for Education, Division of Behavioral and Social Sciences and Education. Washington, DC: The National Academies Press.

National Research Council. (2008). *International Collaborations in Behavioral and Social Sciences Research: Report of a Workshop*. Committee on International Collaborations in Social and Behavioral Sciences Research, U.S. National Committee for the International Union of Psychological Science. Board on International Scientific Organizations and Policy and Global Affairs. Washington, DC: The National Academies Press.

National Research Council. (2012a). *Research Universities and the Future of America: Ten Breakthrough Actions Vital to our Nation's Prosperity and Security.* Committee on Research Universities. Board on Higher Education and Workforce. Policy and Global Affairs Washington, DC: The National Academies Press. Available: http://www.nap.edu/catalog.php?record_id=13396 [May 2014].

National Research Council. (2012b). *Discipline-Based Education Research: Understanding and Improving Learning in Undergraduate Science and Engineering.* S.R. Singer, N.R. Nielsen, and H.A. Schweingruber (Eds.). Committee on the Status, Contributions, and Future Directions of Discipline-Based Education Research. Board on Science Education, Division of Behavioral and Social Sciences and Education. Washington, DC: The National Academies Press.

National Research Council. (2012c). *A Framework for K–12 Science Education: Practices, Crosscutting Concepts, and Core Ideas.* Committee on a Conceptual Framework for New K-12 Science Education Standards. Board on Science Education, Division of Behavioral and Social Sciences and Education. Washington, DC: The National Academies Press.

National Research Council. (2013). *New Directions in Assessing Performance Potential of Individual and Groups: Workshop Summary.* R. Pool, Rapporteur. Committee on Measuring Human Capabilities: Performance Potential of Individuals and Collectives. Board on Behavioral, Cognitive, and Sensory Sciences, Division of Behavioral and Social Sciences and Education. Washington, DC: The National Academies Press.

National Research Council. (2014). *Convergence: Facilitating Transdisciplinary Integration of Life Sciences, Physical Sciences, Engineering, and Beyond.* Committee on Key Challenge Areas for Convergence and Health. Board on Life Sciences, Division on Earth and Life Studies. Washington, DC: The National Academies Press.

National Science Foundation. (2011). *National Science Foundation's Merit Review Criteria: Review and Revisions.* Available: http://www.nsf.gov/nsb/publications/2011/meritreview-criteria.pdf [October 2014].

National Science Foundation. (2012). *FY 2013 Budget Request to Congress.* Available: http://www.nsf.gov/about/budget/fy2013/pdf/EntireDocument_fy2013.pdf [October 2014].

National Science Foundation. (2013). *Report to the National Science Board on the National Science Foundation's Merit Review Process, Fiscal Year 2012.* Available: http://www.nsf.gov/nsb/publications/2013/nsb1333.pdf [May 2014].

National Science Foundation. (2014a). *Cyber-Innovation for Sustainability Science and Engineering.* Program Solicitation No. NSF 14-531. Available: http://www.nsf.gov/pubs/2014/nsf14531/nsf14531.pdf [May 2014].

National Science Foundation. (2014b). *Grant Proposal Guide Chapter II: Proposal Preparation Instructions.* Available: http://www.nsf.gov/pubs/policydocs/pappguide/nsf14001/gpg_2.jsp#IIC2fiegrad [October 2014].

Nature. (2007). Editorial: Peer review reviewed. *Nature 449*(Sept. 13):115. Available: http://www.nature.com/nature/journal/v449/n7159/full/449115a.html [October 2014].

Nellis, M.D. (2014). Defining 21st century land-grant universities through cross-disciplinary research. In M. O'Rourke, S. Crowley, S.D. Eigenbrode, and J.D. Wulfhorst (Eds.), *Enhancing Communication and Collaboration in Interdisciplinary Research* (Ch. 15, pp. 315–334). Thousand Oaks, CA: Sage.

Nembhard, I.M., and Edmondson, A.C. (2006). Making it safe: The effects of leader inclusiveness and professional status on psychological safety and improvement efforts in health care teams. *Journal of Organizational Behavior, 27*(7):941–966. Available: http://onlinelibrary.wiley.com/doi/10.1002/job.413/pdf [October 2014].

Nielsen, M. (2012). *Reinventing Discovery: The New Era of Networked Science.* Princeton, NJ: Princeton University Press.

Norman, D.A. (2013). *The Design of Everyday Things: Revised and Expanded*. New York: Basic Books.

Nunamaker, J.F., Dennis, A.R., Valacich, J.S., Vogel, D., and George, J.F. (1991). Electronic meeting systems. *Communications of the ACM, 34*(7):40–61.

Nunamaker, J.F., Briggs, R.O., Mittleman, D.D., Vogel, D.R., and Balthazard, P.A. (1996/97). Lessons from a dozen years of group support systems research: A discussion of lab and field findings. *Journal of Management Information Systems, 13*(3):163–207. Available: http://groupsystems.files.wordpress.com/2007/09/lessons-from-12-years-of-gss-research.pdf [October 2014].

Obeid, J.S., Johnson, L. M., Stallings, S. and Eichmann, D. (2014). Research networking systems: The state of adoption at institutions aiming to augment translational research infrastructure. *Journal of Translational Medicine and Epidemiology, 2*(2):1026. Available: http://www.jscimedcentral.com/TranslationalMedicine/translationalmedicine-spid-collaboration-science-translational-medicine-1026.pdf [April 2015].

Obstfeld, D. (2005). Social networks, the Tertius Iungens orientation, and involvement in innovation. *Administrative Science Quarterly, 50*:100–130. Available: http://asq.sagepub.com/content/50/1/100.full.pdf [October 2014].

O'Donnell, A.M., and Derry, S.J. (2005). Cognitive processes in interdisciplinary groups: Problems and possibilities. In S.J. Derry, C.D. Schunn, and M.A. Gernsbacher (Eds.), *Interdisciplinary Collaboration: An Emerging Cognitive Science* (pp. 51–82). Mahwah, NJ: Lawrence Erlbaum.

OECD. (2011). *Issue Brief: Public Sector Research Funding. OECD Innovation Policy Platform*. Available: http://www.oecd.org/innovation/policyplatform/48136600.pdf [May 2014].

Office of Management and Budget, Executive Office of the President. (2013). *Memorandum to the Heads of Departments and Agencies: Next Steps in the Evidence and Innovation Agenda*. Available: https://www.whitehouse.gov/sites/default/files/omb/memoranda/2013/m-13-17.pdf [April 2015].

Ohland, M.W., Loughry, M.S., Woehr, D.J., Bullard, L.G., Felder, R.M., Finelli, C.J., et al. (2012). The comprehensive scale for self- and peer-evaluation. *Academy of Management Learning and Education, 11*(4):609–630. Available: http://amle.aom.org/content/11/4/609.full.pdf+html [May 2014].

Okhuysen, G., and Bechky, B. (2009). Coordination in organizations: An integrative perspective. *Annals of the Academy of Management, 3*(1):463–502.

O'Leary, M.B., and Cummings, J.N. (2007). The spatial, temporal, and configurational characteristics of geographic dispersion in teams. *MIS Quarterly, 31*(3):433–452.

O'Leary, M.B., and Mortensen, M. (2010). Go (con)figure: Subgroups, imbalance, and isolates in geographically dispersed teams. *Organization Science, 21*(1):115–131.

O'Leary, M.B., Mortensen, M., and Woolley, A.W. (2011). Multiple team membership: A theoretical model of its effects on productivity and learning for individuals and teams. *Academy of Management Review, 36*(3):461–478. Available: http://www18.georgetown.edu/data/people/mbo9/publication-57723.pdf [October 2014].

O'Leary-Kelly, A.M., Martocchio, J.J., and Frink, D.D. (1994). A review of the influence of group goals on group performance. *Academy of Management Journal, 37*(5):1285–1301.

Olson, G.M., and Olson, J.S. (2000). Distance matters. *Human-Computer Interaction, 15*:139–179. Available: http://www.ics.uci.edu/~corps/phaseii/OlsonOlson-Distance Matters-HCIJ.pdf [October 2014].

Olson, J.S., and Olson, G.M. (2014). *Working Together Apart*. San Rafael, CA: Morgan Claypool.

Ommundsen, Y., Lemyre, P.-N., and Abrahamsen, F. (2010). Motivational climate, need satisfaction, regulation of motivation and subjective vitality: A study of young soccer players. *International Journal of Sports Psychology, 41*(3):216–242.

The Open Source Science Project. (2008–2014). *Research Microfunding Platform.* Available: http://www.theopensourcescienceproject.com/microfinance.php [October 2014].

O'Reilly, C.A. III, and Tushman, M.L. (2004). The ambidextrous organization. *Harvard Business Review* (April):74–83.

O'Rourke, M., and Crowley, S.J. (2013). Philosophical intervention and cross-disciplinary science: The story of the Toolbox Project. *Synthese, 190*(11):1937–1954. Available: https://www.msu.edu/~orourk51/860-Phil/Handouts/Readings/ORourkeCrowley-PhilIntervention&CDScience-Synthese-2012-pg.pdf [October 2014].

O'Rourke, M., Crowley, S., Eigenbrode, S.D., and Wulfhorst, J.D. (Eds.). (2014). *Enhancing Communication and Collaboration in Interdisciplinary Research.* Thousand Oaks, CA: Sage.

Owen-Smith, J. (2001). Managing laboratory work through skepticism: Processes of evaluation and control. *American Sociological Review, 66*(3):427–452.

Owen-Smith, J. (2013). *Workplace Design, Collaboration, and Discovery.* Presented at the National Research Council Workshop on Institutional and Organizational Supports for Team Science, Washington, DC. Available: http://sites.nationalacademies.org/DBASSE/BBCSS/DBASSE_085357 [December 2014].

Patient-Centered Outcomes Research Institute. (2014). *Patient-Centered Outcomes Research Institute Seeks Patient, Scientist and Stakeholder Reviewers for Pilot Projects Grants Program.* Available: http://www.pcori.org/2011/patient-centered-outcomes-research-institute-seeks-patient-scientist-and-stakeholder-reviewers-for-pilot-projects-grants-program/ [May 2015].

Pearce, C.L. (2004). The future of leadership: Combining vertical and shared leadership to transform knowledge work. *Academy of Management Executive, 18*(1):47–57.

Pearce, C.L., and Sims, H.P., Jr. (2002). Vertical versus shared leadership as predictors of the effectiveness of change management teams: An examination of aversive, directive, transactional, transformational, and empowering leader behaviors. *Group Dynamics: Theory, Research, and Practice, 6*(2):172–197.

Pelz, D.C., and Andrews, F.M. (1976). *Scientists in Organizations: Productive Climates for Research and Development.* Ann Arbor: University of Michigan Institute for Social Research.

Pentland, A. (2012). The new science of building great teams. *Harvard Business Review* (April). Available: https://hbr.org/2012/04/the-new-science-of-building-great-teams [January 2015].

Perper, T. (1989). The loss of innovation: Peer review of multi- and interdisciplinary research. *Issues in Integrative Studies, 7*:21–56. Available: http://www.units.miamioh.edu/aisorg/pubs/issues/7_perper.pdf [October 2014].

Petsko, G.A. (2009). Big science, little science. *European Molecular Biology Organization Reports, 10*(12):1282.

Pittinsky, T.L., and Simon, S. (2007). Intergroup leadership. *The Leadership Quarterly, 18*(2007):586–605.

Piwowar, H. (2013). Altmetrics: Value all research products. *Nature, 493*(Jan. 10):159. Available: http://www.nature.com/nature/journal/v493/n7431/full/493159a.html [May 2014].

Pizzi, L., Goldfarb, N.I., and Nash, D.B. (2001). Crew resource management and its applications in medicine. In K.G. Shojania, B.W. Duncan, K.M. McDonald, et al. (Eds.), *Making Health Care Safer: A Critical Analysis of Patient Safety Practices* (pp. 501–510). Rockville, MD: Agency for Healthcare Research and Quality.

Ployhart, R.E, and Moliterno, T.P. (2011). Emergence of the human capital resource: A multilevel model. *Academy of Management Review, 36*(1):127–150. Available: http://is.vsfs.cz/el/6410/zima2012/NA_HRM/um/3904027/2011_Ployhart_Moliterno_Emergence ofHumanCapital_MultilevelModel127.full_1_.pdf [October 2014].

Pohl, C. (2011). What is progress in transdisciplinary research? *Futures, 43*(6):618–626.

Polzer, J.T., Crisp, C.B., Jarvenpaa, S.L., and Kim, J.W. (2006). Extending the faultline model to geographically dispersed teams: How co-located subgroups can impair group functioning. *Academy of Management Journal, 49*(4):679–692.

Porter, A.L., and Rafols, I. (2009). Is science becoming more interdisciplinary? Measuring and mapping six research fields over time. *Scientometrics, 81*(3):719–745. Available: http://www.sussex.ac.uk/Users/ir28/docs/porter-rafols-2009.pdf [February 2014].

Porter, A.L., Cohen, A.S., Roessner, J.D., and Perreault, M. (2007). Measuring researcher interdisciplinarity. *Scientometrics, 72*(1):117–147. Available: http://www.idr.gatech.edu/doc/MRI-Scientometrics_2007.pdf [October 2014].

Priem, J. (2013). Scholarship: Beyond the paper. *Nature, 495*:437–440. Available: http://www.nature.com/nature/journal/v495/n7442/full/495437a.html [May 2014].

Priem, J., Taraborelli, D., Groth, P., and Neylon, C. (2010). *Altmetrics: A Manifesto.* Available: http://altmetrics.org/manifesto/ [May 2015].

Pritchard, R.D., Jones, S.D., Roth, P.L., Stuebing, K.K., and Ekeberg, S.E. (1988). Effects of group feedback, goal setting, and incentives organizational productivity. *Journal of Applied Psychology, 73*(2):337–358.

Pritchard, R.D., Harrell, M.M., DiazGranados, D., and Guzman, M.J. (2008). The productivity measurement and enhancement system: A meta-analysis. *Journal of Applied Psychology, 93*(3):540–567.

Rajivan, P., Janssen, M.A., and Cooke, N.J. (2013). Agent-based model of a cyber-security defense analyst team. *Proceedings of the Human Factors and Ergonomics Society Annual Meeting, 57*(1):314–318.

Rashid, M., Wineman, J., and Zimring, C. (2009). Space, behavior, and environmental perception in open-plan offices: A prospective study. *Environment and Planning B Planning and Design, 36*(3):432–449.

Reid, R.S., Nkedianye, D., Said, M.Y., Kaelo, D., Neselle, M., Makui, O., et al. (2009). Evolution of models to support community and policy action with science: Balancing pastoral livelihoods and wildlife conservation in savannas of East Africa. *Proceedings of the National Academy of Sciences.* Available: http://www.pnas.org/content/early/2009/11/02/0900313106.full.pdf [May 2015].

Rentch, J.R. (1990). Climate and culture: Interaction and qualitative differences in organizational meanings. *Journal of Applied Psychology, 75*:668–681.

Rentsch, J.R., Delise, L.A., Salas, E., and Letsky, M.P. (2010). Facilitating knowledge building in teams: Can a new team training strategy help? *Small Group Research, 41*(5):505–523. Available: http://sgr.sagepub.com/content/41/5/505.full.pdf [October 2014].

Rentsch, J.R., Delise, L.A., Mello, A.L., and Staniewicz, M.J. (2014). The integrative team knowledge building strategy in distributed problem-solving teams. *Small Group Research, 45*(5):568–591. Available: http://sgr.sagepub.com/content/45/5/568.full.pdf [October 2014].

Repko, A.F. (2011). *Interdisciplinary Research: Process and Theory* (2nd ed.). New York: Sage.

Rico, R., Sanchez-Manzanares, M., Antino, M., and Lau, D. (2012). Bridging team faultlines by combining task role assignment and goal structure strategies. *Journal of Applied Psychology, 97*(2):407–420.

Ridgeway, C. (1991). The social construction of status value: Gender and other nominal characteristics. *Social Forces, 70*(2):367–386.

Rijnsoever, F.J., and Hessels, L.K. (2011). Factors associated with disciplinary and interdisciplinary research collaboration. *Research Policy, 40*(3):463–472.

Rosenfeld, P.L. (1992). The potential of transdisciplinary research for sustaining and extending linkages between the health and social sciences. *Social Science and Medicine, 35*(11):1343–1357.

Rubleske, J., and Berente, N. (2012). *Foregrounding the Cyberinfrastructure Center as Cyberinfrastructure Steward.* Presented at the Fifth Annual Workshop on Data-Intensive Collaboration in Science and Engineering, February 11, Bellevue, WA.

Sackett, P.R., Zedeck, S., and Fogli, L. (1988). Relations between measures of typical and maximum performance. *Journal of Applied Psychology, 73*(3):482–486.

Sailer, K., and McColloh, I. (2012). Social networks and spatial configuration—How office layouts drive social interaction. *Social Networks, 34*(1):47–58.

Salas, E., and Lacerenza, D. (2013). *Team Training for Team Science: Improving Interdisciplinary Collaboration.* Presented at the Workshop on Science Team Dynamics and Effectiveness, July 1, National Research Council, Washington, DC. Available: http://sites.nationalacademies.org/DBASSE/BBCSS/DBASSE_083679 [May 2014].

Salas, E., Rozell, D., Mullen, B., and Driskell, J.E. (1999). The effect of team building on performance: Integration. *Small Group Research, 30*(3):309–329. Available: http://sgr.sagepub.com/content/30/3/309.full.pdf [October 2014].

Salas, E., Sims, D.E., and Burke, C.S. (2005). Is there a "Big Five" in teamwork? *Small Group Research, 36*(5):555–599. Available: http://sgr.sagepub.com/content/36/5/555.full.pdf [October 2014].

Salas, E., Cooke, N.J., and Rosen, M.A. (2008). On teams, teamwork, and team performance: Discoveries and developments. *Human Factors: The Journal of the Human Factors and Ergonomics Society, 50*(3):540–547. Available: http://www.ise.ncsu.edu/nsf_itr/794B/papers/Salas_etal_2008_HF.pdf [October 2014].

Salas, E., Goodwin, G.F., and Burke, C.S. (2009). *Team Effectiveness in Complex Organizations: Cross-disciplinary Perspectives and Approaches.* The Organizational Frontiers Series. New York: Routledge/Taylor and Francis Group.

Salas, E., Cooke, N.J., and Gorman, J.C. (2010). The science of team performance: Progress and the need for more... *Human Factors: The Journal of the Human Factors and Ergonomics Society, 52*(2):344–346. Available: http://hfs.sagepub.com/content/early/2010/07/21/0018720810374614.citation [May 2014].

Salazar, M.R., Lant, T.K., and Kane, A. (2011). To join or not to join: An investigation of individual facilitators and inhibitors of medical faculty participation in interdisciplinary research teams. *Clinical and Translational Science, 4*(4):274–278.

Salazar, M.R., Lant, T.K., Fiore, S.M., and Salas, E. (2012). Facilitating innovation in diverse science teams through integrative capacity. *Small Group Research, 43*(5):527–558. Available: http://sgr.sagepub.com/content/43/5/527.full.pdf [October 2014].

Sample, I. (2013). Nobel winner declares boycott of top science journals. *The Guardian,* December 9. Available: http://www.theguardian.com/science/2013/dec/09/nobel-winner-boycott-science-journals [May 2014].

Sarma, A., Redmiles, D., and van der Hoek, A. (2010). Categorizing the spectrum of coordination technology. *IEEE Computer, 43*(6):61–67.

Satzinger, J., and Olfman, L. (1992). A research program to assess user perceptions of group work support. In *Proceedings of the ACM CHI 92 Human Factors in Computing Systems Conference* (pp. 99–106). New York: ACM.

Schaubroeck, J.M., Hannah, S.T., Avolio, B.J., Kozlowski, S.W.J., Lord, R., Trevino, L.K., Peng, C., and Dimotakis, N. (2012). Embedding ethical leadership within and across organizational levels. *Academy of Management Journal, 55*(5):1053–1078. Available: http://psych.wfu.edu/sisr/Articles/Schaubroeck,%20Hannah,%20et%20al.%20(2012).pdf [October 2014].

Schiflett, S.G., Elliott, L.R., Salas, E., and Coovert, M.D. (Eds.). (2004). *Scaled Worlds: Development, Validation and Applications*. London: Hants.

Schnapp, L.M., Rotschy, L., Hall, T.E., Crowley, S., and O'Rourke, M. (2012). How to talk to strangers: Facilitating knowledge sharing within translational health teams with the Toolbox dialogue method. *Translational and Behavioral Medicine*, 2(4):469–479.

Schneider, B., and Barbera, K.M. (Eds.). (2013). *The Oxford Handbook of Organizational Culture and Climate*. Cheltenham, UK: Oxford University Press.

Schneider, B., and Reichers, A.E. (1983). On the etiology of climates. *Personnel Psychology*, 36(1):19–39.

Schneider, B., Wheeler, J.K., and Cox, J.F. (1992). A passion for service: Using content analysis to explicate service climate themes. *Journal of Applied Psychology*, 77(5):705–716.

Schvaneveldt, R.W. (1990). *Pathfinder Associative Networks: Studies in Knowledge Organization*. Norwood, NJ: Ablex.

Scriven, M. (1967). The methodology of evaluation. In R.E. Stake (Ed.), *Curriculum Evaluation*, American Educational Research Association (monograph series on evaluation, no. 1). Chicago: Rand McNally.

Shaman, J., Solomon, S., Colwell, R.R., and Field, C.B. (2013). Fostering advances in interdisciplinary climate science. *Proceedings of the National Academy of Sciences*, 110(Suppl. 1):3653–3656. Available: http://www.pnas.org/content/110/Supplement_1/3653.full. pdf+html [April 2014].

Shrum, W., Genuth, J., and Chompalov, I. (2007). *Structures of Scientific Collaboration*. Cambridge, MA: MIT Press.

Shuffler, M.L., DiazGranados, D., and Salas, E. (2011). There's a science for that: Team development interventions in organizations. *Current Directions in Psychological Science*, 20(6):365–372.

Simons, R. (1995). *Levers of Control: How Managers Use Innovative Control Systems to Drive Strategic Renewal*. Cambridge, MA: Harvard Business School Press.

Simonton, D.K. (2004). *Creativity in Science: Chance, Logic, Genius, and Zeitgeist*. Cambridge, UK: Cambridge University Press.

Simonton, D.K. (2013). Presidential leadership. In M. Rumsey (Ed.), *Oxford Handbook of Leadership* (pp. 327–342). New York: Oxford University Press.

Smith, M.J., Weinberger, C., Bruna, E.M., and Allesina, S. (2014). The scientific impact of nations: Journal placement and citation performance. *PLOS One*, October 8. Available: http://journals.plos.org/plosone/article?id=10.1371/journal.pone.0109195 [January 2015].

Smith-Jentsch, K.A., Zeisig, R.L., Acton, B., and McPherson, J.A. (1998). Team dimensional training: A strategy for guided team self-correction. In J.A. Cannon-Bowers and E. Salas (Eds.), *Making Decisions under Stress: Implications for Individual and Team Training* (pp. 271–297). Washington, DC: American Psychological Association.

Smith-Jentsch, K.A., Milanovich, D.M., and Merket, D.C. (2001). *Guided Team Self-correction: A Field Validation Study*. Presented at the 16th Annual Conference of the Society for Industrial and Organization Psychology, San Diego, CA.

Smith-Jentsch, K.A., Cannon-Bowers, J.A., Tannenbaum, S.I., and Salas, E. (2008). Guided team self-correction: Impacts on team mental models, processes and effectiveness. *Small Group Research*, 39(3):303–327. Available: http://sgr.sagepub.com/content/39/3/303. full.pdf [October 2014].

Smith-Jentsch, K.A., Kraiger, K., Cannon-Bowers, J.A., and Salas, E. (2009). Do familiar teammates request and accept more backup? Transactive memory in air traffic control. *Human Factors*, 51(2):181–192.

Snow, C., Snell, S., Davison, S., and Hambrick, D. (1996). Use transnational teams to globalize your company. *Organizational Dynamics* (Spring):50–67.

Sommerville, M.A., and Rapport, D.J. (2002). *Transdisciplinarity: Recreating Integrated Knowledge*. Montreal, Canada: McGill-Queens University Press.

Sonnenwald, D.H. (2007). Scientific collaboration. *Annual Review of Information Science and Technology, 41*(1):643–681.

Spaapen, J., and Dijstebloem, H. (2005). *Evaluating Research in Context*. The Hague, Netherlands: Consultative Committee of Sector Councils for Research and Development.

Sproull, L., and Kiesler, S. (1991). *Connections: New Ways of Working in the Networked Organization*. Cambridge, MA: MIT Press.

Squier, E., and Davis, E. (1848). *Ancient Monuments of the Mississippi Valley*. Smithsonian Institution Contributions to Knowledge, Vol. 1, Washington, DC.

Stajkovic, A.D., and Luthans, F. (1998). Self-efficacy and work-related performance: A meta-analysis. *Psychological Bulletin, 124*(2):240–261.

Stasser, G., Stewart, D.D., and Wittenbaum, G.M. (1995). Expert roles and information exchange during discussion: The importance of knowing who knows what. *Journal of Experimental Social Psychology, 31*(3):244–265.

Steele, F. (1986). *Making and Managing High Quality Workplaces: An Organizational Ecology*. New York: Teachers College Press.

Steiner, I.D. (1972). *Group Process and Productivity*. New York: Academic Press.

Stevens, M.J., and Campion, M.A. (1994). The knowledge, skill, and ability requirements for teamwork: Implications for human resource management. *Journal of Management, 20*:503–530. Available: http://www.krannert.purdue.edu/faculty/campionm/Knowledge_Skill_Ability.pdf [October 2014].

Stevens, M.J., and Campion, M.A. (1999). Staffing work teams: Development and validation of a selection test for teamwork settings. *Journal of Management, 25*:207–228. Available: http://jom.sagepub.com/content/25/2/207.full.pdf [October 2014].

Stewart, G.L. (2006). A meta-analytic review of relationships between team design features and team performance. *Journal of Management, 32*(1):29–55.

Stipelman, B.A., Feng, A., Hall, K.A., Stokols, D., Moser R.P., Berger, N.A., et al. (2010). *The Relationship Between Collaborative Readiness and Scientific Productivity in the Transdisciplinary Research on Energetics and Cancer (TREC) Centers*. Presented at the 31st Annual Meeting of the Society of Behavioral Medicine, Seattle, WA.

Stipelman, B.A., Hall, K.L., Zoss, A., Okamoto, J., Stokols, D., and Börrner, K. (2014). Mapping the impact of transdisciplinary research: A visual comparison of investigator-initiated and team-based tobacco use research publications. *SciMed Central*, Special Issue on Collaboration Science and Translational Medicine. Available: https://webfiles.uci.edu/dstokols/Pubs/Stipelman.11.translationalmedicine-spid-collaboration-science-translational-medicine-1033.pdf [October 2014].

Stokols, D. (2006). Toward a science of transdisciplinary action research. *American Journal of Community Psychology, 38*:63–77. Available: https://webfiles.uci.edu/psbstaff/1-25-10%20Colloquium-SE%20Perspectives%20on%20Psych%20Research/Toward%20a%20Science%20of%20TD%20Action%20Research_Stokols.pdf [October 2014].

Stokols, D. (2013). *Methods and Tools for Strategic Team Science*. Presented at the Planning Meeting on Interdisciplinary Science Teams, January 11, National Research Council, Washington, DC. Available: http://tvworldwide.com/events/nas/130111/ [May 2014].

Stokols, D. (2014). Training the next generation of transdisciplinarians. In M.O. O'Rourke, S. Crowley, S.D. Eigenbrode, and J.D. Wulfhorst. (Eds.), *Enhancing Communication and Collaboration in Interdisciplinary Research* (pp. 56–81). Thousand Oaks, CA: Sage. Available: https://webfiles.uci.edu/dstokols/Pubs/Stokols%20(2014)%20Training%20the%20next%20generation.pdf [October 2014].

Stokols, D., Fuqua, J., Gress, J., Harvey, R., Phillips, K., Baezconde-Garbanati, L., et al. (2003). Evauating transdisciplinary science. *Nicotine & Tobacco Research*, 5(S-1):S21–S39. Available: https://webfiles.uci.edu/dstokols/Pubs/Stokols%20et%20al%20Eval%20 TD%20Sci.pdf [October 2014].

Stokols, D., Hall, K., Taylor, B., and Moser, R. (2008a). The science of team science: Overview of the field and introduction of the supplement. *American Journal of Preventive Medicine*, 35(Suppl. 2):S77–S89.

Stokols, D., Misra, S., Moser, R.P., Hall, K.L., and Taylor, B.K. (2008b). The ecology of team science: Understanding contextual influences on transdisciplinary collaboration. *American Journal of Preventive Medicine*, 35(2S):S96–S115. Available: http://www.ncbi.nlm. nih.gov/pubmed/18619410 [March 2015].

Stokols, D., Hall, K.L., Moser, R., Feng, A., Misra, S., and Taylor, B. (2010). Cross-disciplinary team science initiatives: Research, training, and translation. In R. Frodeman, J.T. Klein, and C. Mitcham (Eds.), *Oxford Handbook on Interdisciplinarity* (pp. 471–493). Oxford, UK: Oxford University Press.

Stokols, D., Hall, K.L., and Vogel, A.L. (2013). Transdisciplinary public health: Core characteristics, definitions, and strategies for success. In D. Haire-Joshu and T.D. McBride (Eds.), *Transdisciplinary Public Health: Research, Methods, and Practice* (pp. 3–30). San Francisco: Jossey-Bass.

Stout, R.J., Cannon-Bowers, J.A., Salas, E., and Milanovich, D.M. (1999). Planning, shared mental models, and coordinated performance: An empirical link is established. *Human Factors*, 41:61–71.

Strobel, J., and van Barneveld, A. (2009). When is PBL more effective? A meta-synthesis of meta-analyses comparing PBL to conventional classrooms. *Interdisciplinary Journal of Problem-based Learning*, 3(1):43–58. Available: http://docs.lib.purdue.edu/cgi/view content.cgi?article=1046&context=ijpbl [May 2014].

Stvilia, B., Hinnant, C.C., Schindler, K., Worrall, A., Burnett, G., Burnett, K., et al. (2010). *Composition of Science Teams and Publication Productivity.* Presented at ASIST 2010, October 22–27, Pittsburgh, PA.

Surowiecki, J. (2005). *The Wisdom of Crowds.* New York: Anchor.

Swaab, R.I., Schaerer, M., Anicich, E. M., Ronay, R., and Galinsky, A.D. (2014). The too-much-talent effect: Team interdependence determines when more talent is too much or not enough. *Psychological Science*, 25:1581–1591. Available: http://www8.gsb.columbia. edu/cbs-directory/sites/cbs-directory/files/publications/Too%20much%20talent%20PS. pdf [October 2014].

Swezey, R., and Salas, E. (Eds.). (1992). *Teams: Their Training and Performance.* Norwood, NJ: Ablex.

Taggar, S., and Brown, T.C. (2001). Problem-solving team behaviors: Development and validation of BOS and a hierarchical factor structure. *Small Group Research*, 32:698–726. Available: http://sgr.sagepub.com/content/32/6/698.full.pdf [October 2014].

Tajfel, H. (1982). Social psychology of intergroup relations. *Annual Review of Psychology*, 33:1–39.

Tajfel, H., and Turner, J C. (1986). The social identity theory of intergroup behaviour. In S. Worchel and W.G. Austin (Eds.), *Psychology of Intergroup Relations* (pp. 7–24). Chicago: Nelson-Hall.

Takeuchi, K., Lin, J.C., Chen, Y., and Finholt, T. (2010). Scheduling with package auctions. *Experimental Economics*, 13(4):476–499.

Tang, J.C., Zhao, C., Xiang, C., and Inkpen, K. (2011). Your time zone or mine?: A study of globally time zone-shifted collaboration. In *Proceedings of the ACM 2011 Conference on Computer Supported Cooperative Work* (pp. 235–244). New York: ACM. Available: http://www.msr-waypoint.com/pubs/143832/pr295-tang.pdf [June 2014].

Tannenbaum, S.I., Mathieu, J.E., Salas, E., and Cohen, D. (2012). Teams are changing: Are research and practice evolving fast enough? *Industrial and Organizational Psychology,* 5(1):2–24.

Teasley, R.W., and Robinson, R.B. (2005). Modeling knowledge-based entrepreneurship and innovation in Japanese organizations. *International Journal of Entrepreneurship,* 9:19–44.

Thatcher, S.M.B., and Patel, P.C. (2011). Demographic faultlines: A meta-analysis of the literature. *Journal of Applied Psychology, 96(6):*1119–1139.

Thatcher, S.M.B., and Patel, P.C. (2012). Group faultlines a review, integration, and guide to future research. *Journal of Management, 38(4):*969–1009. Available: http://jom.sagepub.com/content/38/4/969.full.pdf [October 2014].

Thompson, B.M., Schneider, V.F., Haidet, P., Levine, R.E., McMahon, K.K., Perkowski, L.C., and Richards, B.F. (2007). Team-based learning at ten medical schools: Two years later. *Medical Education, 41:*250–257. Available: http://medicina.fm.usp.br/cedem/did/seminarios/TBI_ten_med_school_2y_later_MedEduc2007.pdf [October 2014].

Toker, U., and Gray, D.O. (2008). Innovation spaces: Workspace planning and innovation in U.S. university research centers. *Research Policy, (37)2:*309–329.

Toubia, O. (2006). Idea generation, creativity, and incentives. *Marketing Science,* 25:411–425.

Traweek, S. (1988). *Beamtimes and Lifetimes: The World of High-Energy Physicists.* Cambridge, MA: Harvard University Press.

Trochim, W., Marcus, S., Masse, L., Moser, R., and Weld, P. (2008). The evaluation of large research initiatives—A participatory integrative mixed-methods approach. *American Journal of Evaluation,* 29(1):8–28. Available: http://www.socialresearchmethods.net/research/eli.pdf [October 2014]

Tscharntke, T., Hochberg, M.E., Rand, T.A., Resh, V.H., and Krauss, J. (2007). Author sequence and credit for contributions in multi-authored publications. *PLoS Biology,* 5(1):e18. Available: http://www.plosbiology.org/article/fetchObject.action?uri=info%3Adoi%2F10.1371%2Fjournal.pbio.0050018&representation=PDF [May 2014].

Tuckman, B.W. (1965). Developmental sequence in small groups. *Psychological Bulletin,* 63(6):384–399.

Uhl-Bien, M., and Pillai, R. (2007). The romance of leadership and the social construction of followership. In B. Shamir, R. Pillai, M. Bligh, and M. Uhl-Bien (Eds.), *Follower-Centered Perspectives on Leadership: A Tribute to the Memory of James R. Meindl* (pp. 187–210). Charlotte, NC: Information Age.

Uhl-Bien, M., Riggio, R.E., Lowe, K.B., and Carsten, M.K. (2014). Followership theory: A review and research agenda. *The Leadership Quarterly,* 25:83–104.

University of Southern California. (2011). *Guidelines for Assigning Authorship and Attributing Contributions to Research Products and Creative Works.* Los Angeles: Author. Available: https://research.usc.edu/files/2011/07/URC_on_Authorship_and_Attribution_10.20111.pdf [April 2015].

U.S. Department of Education. (2013). *Digest of Education Statistics, 2012.* Washington, DC: National Center for Education Statistics. Available: https://nces.ed.gov/programs/digest/d12/index.asp [April 2015].

U.S. Government Accountability Office. (2012). *James Webb Space Telescope: Actions Needed to Improve Cost Estimate and Oversight of Test and Integration.* Washington, DC: Author. Available: http://www.gao.gov/assets/660/650478.pdf [September 2014].

Uzzi, B., Mukerjee, S., Stringer, M., and Jones, B.F. (2013). Atypical combinations and scientific impact. *Science,* 342(6157):468–472. Available: http://www.kellogg.northwestern.edu/faculty/uzzi/htm/papers/Science-2013-Uzzi-468-72.pdf [October 2014].

Van der Vegt, G.S., and Bunderson, J.S. (2005). Learning and performance in multidisciplinary teams: The importance of collective team identification. *Academy of Management Journal,* 48(3):532–547.

Van Ginkel, W., Tindale, R.S., and van Knippenberg, D. (2009). Team reflexivity, development of shared task representations, and the use of distributed information in group decision making. *Group Dynamics: Theory, Research, and Practice*, 13(4):265–280.

Van Knippenberg, D., De Dreu, C.K.W., and Homan, A.C. (2004). Work group diversity and group performance: An integrative model and research agenda. *Journal of Applied Psychology*, 89(6):1008–1022.

VandeWalle, D. (1997). Development and validation of a work domain goal orientation instrument. *Educational and Psychological Measurement*, 57(6):995–1015. Available: http://dvandewalle.cox.smu.edu/goal_orient_instrum.pdf [October 2014].

Vermeulen, N., Parker, J.N., and Penders, B. (2010). Big, small, or mezzo?: Lessons from science studies for the ongoing debate about 'big' versus 'little' research projects. *EMBO Reports*, 11(6):420–423. Available: http://onlinelibrary.wiley.com/doi/10.1038/embor.2010.67/pdf [October 2014].

Vincente, K.J. (1999). *Cognitive Work Analysis: Toward Safe, Productive, and Healthy Computer-Based Work*. Mahwah, NJ: Lawrence Erlbaum.

Vinokur-Kaplan, D. (1995). Treatment teams that work (and those that don't) application of Hackman's Group Effectiveness Model to interdisciplinary teams in psychiatric hospitals. *Journal of Applied Behavioral Science*, 31(3):303–327.

Vogel, A., Feng, A., Oh, A., Hall, K.L., Stipelman, B. A., Stokols, D., et al. (2012). Influence of a National Cancer Institute transdisciplinary research and training initiative on trainees' transdisciplinary research competencies and scholarly productivity. *Translational Behavioral Medicine*, 2(4):459–468. Available: http://www.ncbi.nlm.nih.gov/pubmed/24073146 [May 2014].

Vogel, A.L., Stipelman, B.A., Hall, K.L., Nebeling, D., Stokols, D., and Spruijt-Metz, D. (2014). Pioneering the transdisciplinary team science approach: Lessons learned from National Cancer Institute grantees. *Journal of Translational Medicine & Epidemiology*, 2(2):1027. Available: http://www.jscimedcentral.com/TranslationalMedicine/translationalmedicine-spid-collaboration-science-translational-medicine-1027.pdf [September 2014].

Voida, A., Olson, J.S., and Olson, G.M. (2013). Turbulence in the clouds: Challenges of cloud-based information work. In *Proceedings of the SIGCHI Conference on Human Factors in Computing Systems* (pp. 2273–2282). New York: ACM.

Volpe, C.E., Cannon-Bowers, J.A., Salas, E., and Spector, P. (1996). The impact of cross-training on team functioning: An empirical investigation. *The Journal of the Human Factors and Ergonomics Society*, 38(1):87–100. Available: http://hfs.sagepub.com/content/38/1/87.full.pdf [October 2014].

Wagner, C.S., Roessner, J.D., Bobb, K., Klein, J.T., Boyack, K.W., Keyton, J., et al. (2011). Approaches to understanding and measuring interdisciplinary scientific research (IDR): A review of the literature. *Journal of Informetrics* 5(1):14–26.

Wang, D., Waldman, D.A., and Zhang, Z. (2014). A meta-analysis of shared leadership and team effectiveness. *Journal of Applied Psychology*, 99(2):181–198.

Wegner, D.M. (1995). A computer network model of human transactive memory. *Social Cognition*, 13(3):319–339.

Wegner, D.M., Giuliano, T., and Hertel, P. (1985). Cognitive interdependence in close relationships. In W.J. Ickes (Ed.), *Compatible and Incompatible Relationships* (pp. 253–276). New York: Springer-Verlag.

Westfall, C. (2003). Rethinking big science: Modest, mezzo, grand science and the development of the Bevelac, 1971–1993. *Isis*, 94:30–56.

Whittaker, S. (2013). *Collaboration Technologies: Response*. Presented at the National Research Council's Workshop on Institutional and Organizational Supports for Team Science, October 24, Washington, DC. Available: http://sites.nationalacademies.org/DBASSE/BBCSS/DBASSE_085357 [April 2014].

Whittaker, S., and Sidner, C. (1996). *E-mail Overload: Exploring Personal Information Management of E-mail.* Presented at the CHI '96, Vancouver, BC, Canada. Available: https://www.ischool.utexas.edu/~i385q/readings/Whittaker_Sidner-1996-Email.pdf [October 2014].

Whittaker, S., Bellotti, V., and Moody, P. (2005). Introduction to this special issue on revisiting and reinventing e-mail. *Human-Computer Interaction,* 20(1–2):1–9.

Wickens, C.D, Lee J.D., Liu, Y., and Gorden Becker, S.E. (1997). *An Introduction to Human Factors Engineering* (2nd ed.). Englewood Cliffs, NJ: Prentice Hall.

Wiersema, M.F., and Bird, A. (1993). Organizational demography in Japanese firms: Group heterogeneity, individual dissimilarity, and top management team turnover. *Academy of Management,* 36(5):996–1025.

Willey, G., Smith, A., Tourtellot III, G., and Graham, I. (1975). *Excavations at Seibal, Guatemala; Introduction: A Site and Its Setting.* Memoirs of the Peabody Museum, Harvard University, Vol. 13, No. 1. Cambridge, MA: Harvard University Press.

Winter, S.J., and Berente, N. (2012). A commentary on the pluralistic goals, logics of action, and institutional contexts of translational team science. *Translational Behavioral Medicine,* 2(4):441–445.

Woolley, A.W., Gerbasi, M.E., Chabris, C.F., Kosslyn, S.M., and Hackman, J.R. (2008). Bringing in the experts: How team composition and collaborative planning jointly shape analytic effectiveness. *Small Group Research,* 39(3):352–371.

Woolley, A.W., Chabris, C.F., Pentland, A., Hashmi, N., and Malone, T.W. (2010). Evidence for a collective intelligence factor in the performance of human groups. *Science,* 330:686–688. Available: http://www.sciencemag.org/content/330/6004/686 [December 2014].

Wu, J.B., Tsui, A.S., and Kinicki, A.J. (2010). Consequences of differentiated leadership in groups. *Academy of Management Journal,* 53(1):90–106.

Wuchty, S., Jones, B.F., and Uzzi, B. (2007). The increasing dominance of teams in production of knowledge. *Science,* 316:1036–1038. Available: http://www.kellogg.northwestern.edu/faculty/jones-ben/htm/teams.printversion.pdf [October 2014].

Wulf, W.A. (1993). The collaboratory opportunity. *Science,* 261(5123):854–855.

Zaccaro, S.J., and DeChurch, L.A. (2012). Leadership forms and functions in multiteam systems. In S.L. Zaccaro, M.A. Marks, and L.A. DeChurch (Eds.), *Multiteam Systems: An Organization Form for Dynamic and Complex Environments* (pp. 253–288). New York: Routledge.

Zerhouni, E.A. (2005). Translational and clinical science—time for a new vision. *New England Journal of Medicine,* 353(15):1621–1623.

Zheng, J.B., Veinott, E., Bos, N., Olson, J.S., and Olson, G.M. (2002). Trust without touch: Jumpstarting long-distance trust with initial social activities. In *CHI 2002 Proceedings of the SIGCHI Conference on Human Factors in Computing Systems* (pp.141–146). New York: ACM. Available: http://dl.acm.org/citation.cfm?id=634241[June 2014].

Zohar, D. (2000). A group-level model of safety climate: Testing the effect of group climate on microaccidents in manufacturing jobs. *Journal of Applied Psychology,* 85(4):587–596.

Zohar, D. (2002). Modifying supervisory practices to improve subunit safety: A leadership-based intervention model. *Journal of Applied Psychology,* 87(1):156–163.

Zohar, D., and Hofmann, D.A. (2012). Organizational culture and climate. In S.W.J. Kozlowski (Ed.), *The Oxford Handbook of Organizational Psychology, Volume 1.* Cheltenham, UK: Oxford University Press.

Zohar, D., and Luria, G. (2004). Climate as a social-cognitive construction of supervisory safety practices: Scripts as proxy of behavior patterns. *Journal of Applied Psychology,* 89(2):322–333.

Appendix

Biographical Sketches of Committee Members

NANCY J. COOKE (*Chair*) is a professor of applied psychology at Arizona State University and science director and on the board of directors of the Cognitive Engineering Research Institute in Mesa, Arizona. She is also a section editor of *Human Factors* and serves on the Air Force Scientific Advisory Board. Currently, she supervises postdoctoral, graduate, and undergraduate research on team cognition with applications in design and training for military command-and-control systems, emergency response, medical systems, and uninhabited aerial systems. Cooke received a B.A. in psychology from George Mason University and received her M.A. and Ph.D. in cognitive psychology in 1983 and 1987, respectively, from New Mexico State University.

ROGER D. BLANDFORD (NAS) is the Luke Blossom professor in the School of Humanities and the Sciences at Stanford University, where he also serves as director of the Kavli Institute for Particle Astrophysics and Cosmology. His research interests cover many aspects of particle astrophysics and cosmology. He was an undergraduate and research student at Cambridge University and held postdoctoral positions at Cambridge University, Princeton University, and the University of California, Berkeley, before joining the faculty of the California Institute of Technology in 1976.

JONATHON N. CUMMINGS is an associate professor of management at the Fuqua School of Business, Duke University. He was an assistant professor in the MIT Sloan School of Management, where he received an National Science Foundation Early Career Award for his research on innovation in geographically dispersed teams and networks. His subsequent research has focused on virtual teams in corporations as well as collaboration in science. He earned his B.A. in psychology from the University of Michigan, A.M. in psychology from Harvard University, and Ph.D. in organization sciences from Carnegie Mellon University.

STEPHEN M. FIORE is an associate professor of cognitive sciences in the University of Central Florida's (UCF) Department of Philosophy and director of the Cognitive Sciences Laboratory at UCF's Institute for Simulation and Training. He also serves as the current president of the Interdisciplinary Network for Group Research and is a founding program committee member for the annual Science of Team Science conference. His primary area of research is the interdisciplinary study of complex collaborative problem solving. He has a Ph.D. in cognitive psychology from the University of Pittsburgh, Learning Research and Development Center.

KARA L. HALL is a health scientist, director of the Science of Team Science Team, and co-director of the Theories Project in the Science of Research and Technology Branch at the National Cancer Institute. She served as a co-chair for the 2006 conference "The Science of Team Science: Assessing the Value of Transdisciplinary Research" and co-edited the *American Journal of Preventive Medicine* Special Supplement on the Science of Team Science. She earned her master's and doctoral degrees in psychology with specializations in clinical psychology, neuropsychology, and behavioral science at the University of Rhode Island.

JAMES S. JACKSON (IOM) is the Daniel Katz distinguished university professor of psychology, professor of health behavior and health education, School of Public Health, and director and research professor of the Institute for Social Research, at the University of Michigan. He is the past chair of the Social Psychology Training Program and director of the Research Center for Group Dynamics, the Program for Research on Black Americans, and the Center for Afroamerican and African Studies, all at the University of Michigan. He earned his Ph.D. in social psychology from Wayne State University.

JOHN L. KING is the W.W. Bishop professor of information, former dean of the School of Information, and former vice provost at the University of Michigan. In 2000, he joined the University of Michigan after 20 years on the faculty of the University of California, Irvine. He has been Marvin Bower fellow at the Harvard Business School, distinguished visiting professor at the National University of Singapore and at Nanyang Technological University in Singapore, and Fulbright distinguished chair in American

Studies at the University of Frankfurt. He holds a Ph.D. in administration from the University of California, Irvine, and an honorary doctorate in economics from Copenhagen Business School.

STEVE W.J. KOZLOWSKI is a professor of organizational psychology at Michigan State University. His research is focused on the design of active learning systems and the use of "synthetic experience" to train adaptive skills, systems for enhancing team learning and team effectiveness, and the critical role of team leaders in the development of adaptive teams. He holds a B.A. in psychology from the University of Rhode Island and an M.S. and a Ph.D. in organizational psychology from Pennsylvania State University.

JUDITH S. OLSON is the Bren professor of information and computer sciences in the Informatics Department at the University of California, Irvine, with courtesy appointments in the School of Social Ecology and the Merage School of Business. She has researched teams whose members are not co-located for more than 20 years. Her current work focuses on ways to verify her theory's components while at the same time helping new scientific collaborations succeed. She has also been studying the adoption of the new suite of collaboration tools in Google Apps. She holds a B.A. from Northwestern University and a Ph.D. from the University of Michigan.

JEREMY A. SABLOFF (NAS) is the president of the Santa Fe Institute. Before joining the institute, he taught at Harvard University, University of Utah, University of New Mexico (where he was chair of the department), University of Pittsburgh (where he also was chair), and University of Pennsylvania (where he was the Williams director of the University of Pennsylvania Museum and Christopher H. Browne distinguished professor of anthropology). He also was an overseas visiting fellow at St. John's College, Cambridge, England. He earned a B.A. from the University of Pennsylvania and a Ph.D. from Harvard University.

DANIEL S. STOKOLS is a research professor and chancellor's professor emeritus in psychology and social behavior and planning, policy, and design at the University of California, Irvine (UCI). He holds courtesy appointments in public health, epidemiology, and nursing sciences at UCI. He served as director and founding dean of the UCI School of Social Ecology from 1988 to 1998. He earned his B.A. at the University of Chicago and his Ph.D. in social psychology at the University of North Carolina at Chapel Hill.

BRIAN UZZI is the Richard L. Thomas distinguished professor of leadership at the Kellogg School of Management, Northwestern University. He also directs the Northwestern University Institute on Complex Systems and is a professor of sociology and management science at the McCormick School of Engineering. He has also been on the faculty of Harvard University, INSEAD, University of Chicago, and University of California, Berkeley, where he was the Warren E. and Carol Spieker professor of leadership. He has a B.A. in business economics from Hofstra University, and a Ph.D. in sociology from State University of New York, Stony Brook.

HANNAH VALANTINE is the chief officer for scientific workforce diversity at the National Institutes of Health and professor of cardiovascular medicine at the Stanford University School of Medicine. Her research interests include diversity of the scientific workforce and pathophysiology of transplant-related atherosclerosis. In November 2004, she was appointed as senior associate dean for diversity and leadership in the Stanford University School of Medicine. She is a graduate of St. George's Hospital, London University. She earned her M.D. from London University, London, completed her residency at St. George's Hospital, Brompton Hospital and Guys Hospital London, and her cardiology fellowship training at Royal Postgraduate Medical School in Hammersmith London.